Algebraic Geometry and Commutative Algebra

Algebraic Geometry and Commutative Algebra

Editor

Linsen Chou

Algebraic Geometry and Commutative Algebra

Edited by **Linsen Chou**

Printed in 2017

ISBN: 978-1-68117-182-1

Library of Congress Control Number: 2015949100

© 2016 by
SCITUS Academics LLC,
616, Corporate Way, Suite 2, 4766,
Valley Cottage, NY 10989

www.scitusacademics.com

Preface

Algebraic geometry is a fascinating branch of mathematics that combines methods from both, algebra and geometry. It transcends the limited scope of pure algebra by means of geometric construction principles. The major objects of study in algebraic geometry are algebraic varieties, which are geometric manifestations of solutions of systems of polynomial equations. Examples of the most studied classes of algebraic varieties are: plane algebraic curves, which include lines, circles, parabolas, ellipses, hyperbolas, cubic curveslike elliptic curves and quartic curves like lemniscates, and Cassini ovals.

Commutative algebra is the branch of algebra that studies commutative rings, theirideals, and modules over such rings. Both algebraic geometry and algebraic number theory build on commutative algebra. Commutative algebra is the main technical tool in the local study of schemes.The study of rings which are not necessarily commutative is known asnoncommutative algebra; it includes ring theory, representation theory, and the theory of Banach algebras.

This textbook provides a gateway into the two difficult fields of algebraic geometry and commutative algebra. Algebraic geometry, supported fundamentally by commutative algebra, is a cornerstone of pure mathematics. This book explores into the rich interplay between algebraic geometry and commutative algebra.

Contents

The Behavior of Normality when Iteratively Finding the Normal to a Line in an l_p Geometry

Joshua M. Fitzhugh and
David L. Farnsworth
School of Mathematical Sciences, Rochester
Institute of Technology, Rochester, USA.

ABSTRACT

The normal direction to the normal direction to a line in Minkowski geometries generally does not give the original line. We show that in l_p geometries with p>1 repeatedly finding the normal line through the origin gives sequences of lines that monotonically approach specific lines of symmetry of the unit circle. Which lines of symmetry that are approached depends upon the value of p and the slope of the initial line.

INTRODUCTION

Minkowski geometries are completely characterized by their unit circle, which is centrally symmetric about the origin and convex [1: p. 17]. The spaces are homogeneous (all points are the same) and generally anisotropic (the yard stick for distance is not the same in all directions). Our principal interest is the planar Minkowski l_p geometries with $p \geq 1$. Their unit circles are

$$|x|^p + |y|^p = 1.$$

(1)

The exponent p must be at least 1 in order for the unit circle to be convex. Convexity is required for the triangle inequality [1: pp. 22, 23]. If p=2, this is Euclidean geometry. If p=1, the circle is not strictly convex. As discussed in Section 2, since in Minkowski geometries a necessary and sufficient condition for uniqueness of normal directions to lines is that the unit circle be strictly convex, we do not consider the p=1 case. Convex unit circles are strictly convex if they contain no line segments. The l1 geometry is well studied and is sometimes called taxicab, Manhattan, or city-block geometry [2]. Since the unit circle for the limiting case $p \to \infty$ is the square with vertices $(\pm 1, \pm 1)$, we do not consider that geometry, as well. Figure 1 shows some l_p unit circles and the circle with p=0.5, which does not produce a Minkowski geometry since the circle is not convex.

The unit circle determines distances. For the Minkowski distance between points P_1 and P_2, consider line L through the origin O and parallel to the line through P_1 and P_2. The distance between P_1 and P_2 is the quotient of the Euclidean distance between P_1 and P_2 and the unit of measurement or scale in the direction of L. The unit of measurement is the Euclidean distance from the O to point Q where L intersects the unit circle [3: p. 225, 4]. Equivalently, translate the axes so that the origin is at P_1 and the point P_2 has coordinates (x_2, y_2). The Minkowski distance between points P_1 and P_2 is the value of d>0 such that $(x_2/d, y_2/d)$, is on the unit circle [1: p. 17]. These definitions give a distance function [1: pp. 17-18, 3: pp. 225-228]. For l_p geometries, the second definition gives

$$\left| x_2 / d \right|^p + \left| y_2 / d \right|^p = 1 \, ,$$

So that

$$d = \left(\left| x_2 \right|^p + \left| y_2 \right|^p \right)^{1/p} .$$

There are many applications of l_p geometries. The shape is called a Lamé curve after some work by Gabriel Lamé. Ruane and Swartzlander [5] considered apertures for light with shape (2) with p>2, which give a larger area than p=2 for their constraints. Piet Hein designed a large traffic island for Stockholm, Sweden using (2) with p=2.5 and a=1.2b, saying that it gives a smooth traffic flow. He called the curves (2) with p>2 super-ellipses. The shape (2) has been extensively used for furniture design and elsewhere [6: pp. 240-254]. The Melior typeface's "O" has p=2.7581, perhaps for aesthetic reasons.

$$\left|x/a\right|^{p} + \left|y/b\right|^{p} = 1 \tag{2}$$

In the next section, we define normality in Minkowski geometries. Since the normal line to the normal line of a line is usually not the original line, in Section 3 we determine the behavior of the lines obtained by successively finding normal lines of normal lines. The limiting behavior is in Theorems 3.4 and 3.5. In Section 4, we create a circle, called a Radon curve, using portions of two l_p geometries' unit circles, for which the normal to the normal of any line is the original line, which is called reflexivity of normality.

DEFINITION OF NORMALITY

There are two equivalent, intuitive ways to define normality in Minkowski geometries with smooth unit circles. One is that line L_2 is normal to the given line L_1 with L_2 meeting L_1 at point Q if for every point P on line L_2, the distance from P to Q is the minimum of all distances from P to any point on L_1 [1: p. 78, 3: p. 228].

For easier expression, we give the other definition in terms of unit vectors. It says that a unit vector is normal to a second unit vector if the first vector contains the origin and a point where the slope of the unit circle is the same as the slope of the second vector [1: p. 125, 7: p. 145]. An application of this definition is illustrated in Figure 2, where 6 p=6. We use the second definition, since in practice finding the tangent lines to (1) is easier than minimizing a distance.

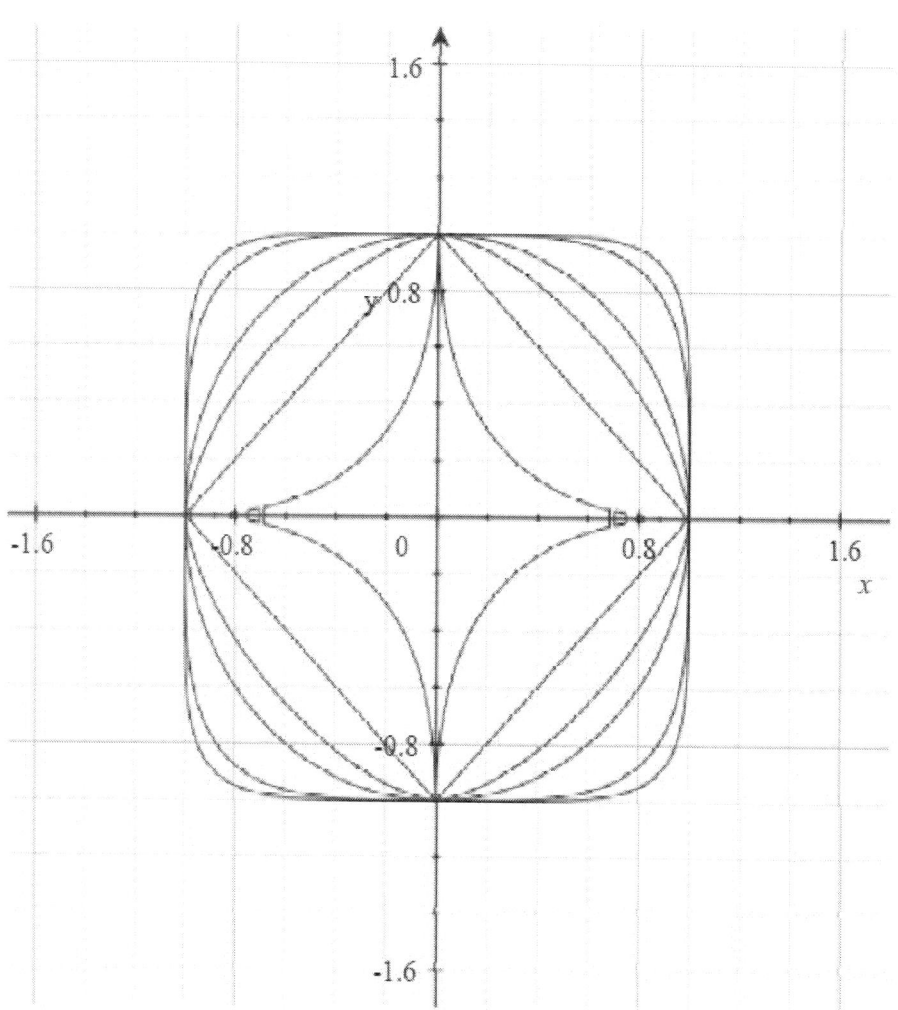

Figure 1: Unit circles $|x|^p + |y|^p = 1$ for p = 0.5, 1, 1.5, 2, 4 and 6. The circles with p = 1, 1.5, 2, 4 and 6 give Minkowski geometries.

In l_p geometries, the axes x=0 and y=0 are mutually normal lines, as are y=x and y=-x. However, in general, the normal line to the normal line of a line is not the original line.

In any Minkowski geometry, the unit circle is strictly convex if and only if normality is unique [1: p. 257, 3: p. 232]. If the unit circle contains

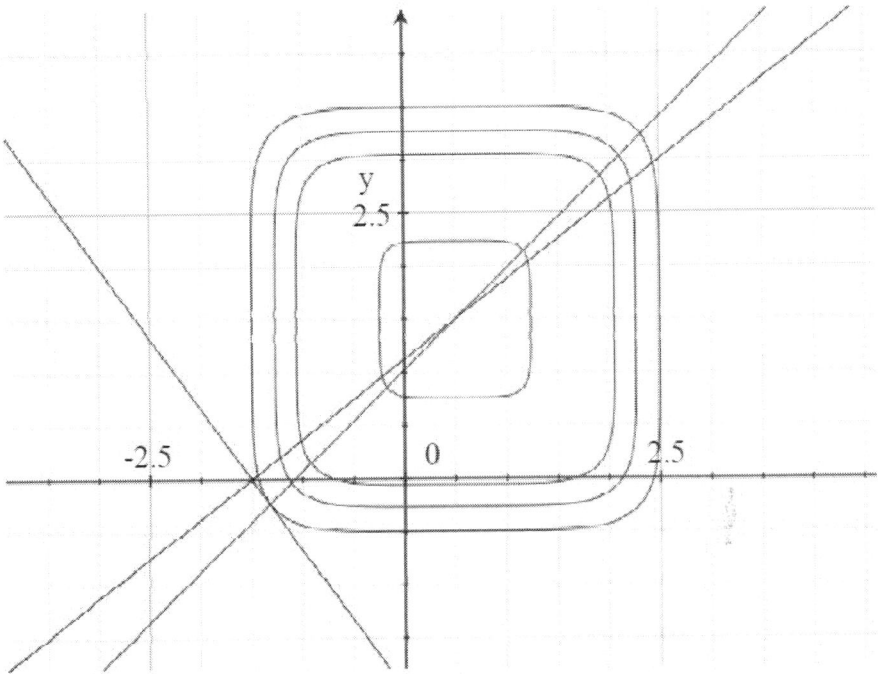

Figure 2: In l6 geometry, the line orthogonal to $y = -(4/3)x - 2$ through the point $(0.5, 1.5)$ is $y = 0.95x + 1.025$, but the Euclidean (l_2) normal line is $y = (3/4)x + 9/8$.

a line segment S, then normality is not unique for any line parallel to that segment. Take such a line L through the origin. Any line through the origin and intersecting S is normal to L, since the distance from the origin to the segment is one for all the normal lines. Hence, we do not consider l_1 or l_∞ geometries.

REPEATEDLY FINDING NORMAL LINES

The purpose of this section is to explore the behavior of the lines found by repeatedly finding normal lines in l_p geometries. The origin O is placed at the point on the initial line where the normal is found.

Lemma 3.1: Consider l_p geometry with p>1. For m>0, the slope of the normal line to y=mx is

$$-(1/m)^{1/(p-1)}.$$

(3)

For m<0, the slope of the normal line to y=mx is

$$(-1/m)^{1/(p-1)}.$$

(4)

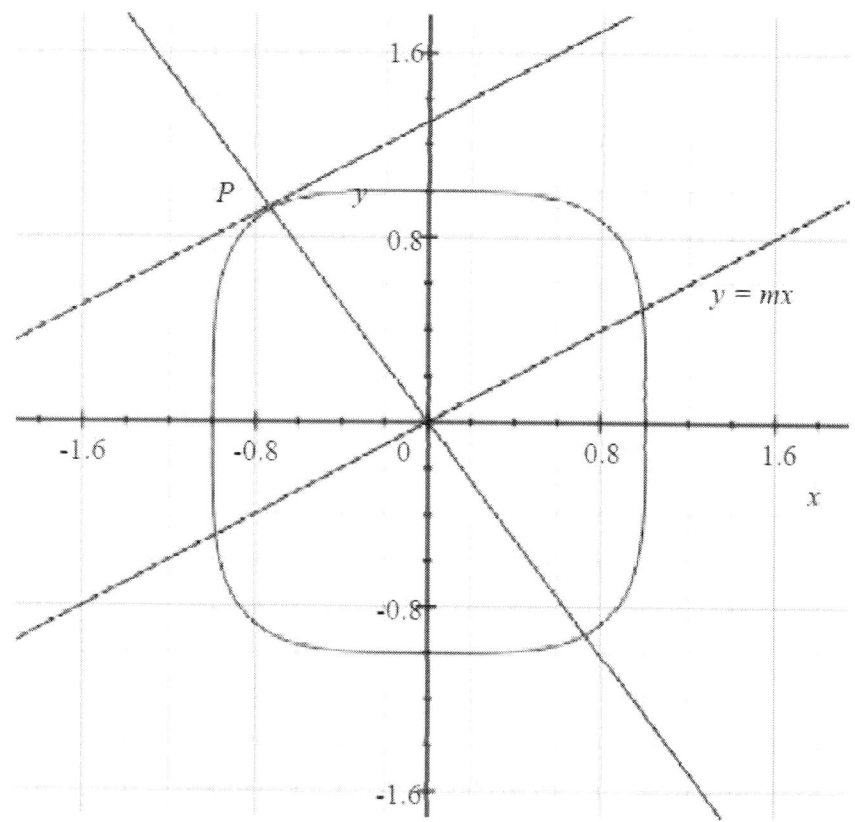

Figure 3: OP is normal to y=mx.

Proof: For m>0, we find the point of tangency to the unit circle, where the tangent is parallel to y=mx. See Figure 3. In the second quadrant, the derivative of $(-x)^p + y^p = 1$ gives $dy/dx = (-x/y)^{p-1}$ setting this equal to m gives $y/x = -(-1/m)^{1/(p-1)}$, which is the slope of the normal line to y=mx Formula (4) is derived similarly.

Lemma 3.2: Consider l_p geometry with p>1 Designated by m_n the slope of the nth line found by iteratively finding normal lines at the origin, starting with the line y=m_0x with m_0>0 For even n,

$$m_{n+2} = (m_n)^{1/(p-1)^2},$$
(5)

And for odd n,

$$m_{n+2} = -(-m_n)^{1/(p-1)^2}.$$
(6)

For even n,

$$m_n = (m_0)^{1/(p-1)^n},$$
(7)

And for odd n,

$$m_n = -(1/m_0)^{1/(p-1)^n}.$$
(8)

These formulas can be appropriately altered for m_0<0.

Proof: To obtain (5), for even n, using (3),

$$m_{n+1} = -(1/m_n)^{1/(p-1)}$$
(9)

and using (4) and (9),

$$m_{n+2} = \left(-1/m_{n+1}\right)^{1/(p-1)}$$

$$= \left[-1/\left\{-\left(1/m_n\right)^{1/(p-1)}\right\}\right]^{1/(p-1)}$$

$$= \left(m_n\right)^{1/(p-1)^2}.$$

Equation (5) supplies $m_2 = (m_0)^{1/(p-1)^2}$ and also the main induction step to give (7). Similarly, obtain (6) and (8).

Lemma 3.3: Consider l_p geometry with p>1. If

$$m_{0.1} = 1/m_{0.2} ,$$

(10)

Then

$$m_{n.1} = 1/m_{n.2} ,$$

(11)

Where the second subscript indicates the identity of the line.

Proof: Take $m_{0,1}$ and $m_{0,2}$ to be positive. The proof for negative initial slopes is similar. For even n, using (7) for line 1, (10), and then (7) for line 2 give

$$m_{n,1} = \left(m_{0,1}\right)^{1/(p-1)^n} = \left(1/m_{0,2}\right)^{1/(p-1)^n} = 1/m_{n,2} .$$

The proof of (11) for odd n uses (8) and (10) in a similar manner.

Because of the symmetries of the l_p unit circle about the axes, only $m_0 > 0$ need be considered. The condition $m_{0,1} = 1/m_{0,2}$ between the slopes of two initial lines means that the lines have the same angle with the respective axes. Lemma 3.3 shows the symmetries about $y = \pm x$ in

The Behavior of Normality when Iteratively Finding the Normal

the behavior of the iterated normal lines, so only initial slopes between 0 and 1 need to be considered.

Theorem 3.4 Consider I_p geometry with p>2. For the initial line y=m_0x with 0<m_0<1, the subsequence of $\{m_n\}$ for even n has values 0<m_n<1 and monotonically approaches 1, and the subsequence for odd n has values m_n<-1 and monotonically approaches −1. For the initial line y=m_0x with m_0>1, the subsequence of $\{m_n\}$ for even n has values m_n>1 and monotonically approaches 1, and the subsequence for odd n has values m_n>-1 and monotonically approaches −1.

Proof: Take 0<m_0<1. Using Lemma 3.2, for even n,

$$m_{n+2} = \left(m_n\right)^{1/(p-1)^2} > m_n$$

And

$$\operatorname*{Limit}_{n \to \infty} m_n = \operatorname*{Limit}_{n \to \infty} \left(m_0\right)^{1/(p-1)^n} = 1.$$

For odd n,

$$m_{n+2} = -\left(-m_n\right)^{1/(p-1)^2} > m_n$$

And

$$\operatorname*{Limit}_{n \to \infty} m_n = \operatorname*{Limit}_{n \to \infty} \left\{-\left(1/m_0\right)^{1/(p-1)^n}\right\} = -1.$$

Lemma 3.3 says that initial lines y=m_0x with 0<m_0<1 give the behavior of the iterated normal lines for m_0>1.

As an example of Theorem 3.4, Table 1 contains the slopes of the first eight iterated normal lines for p=2.5 with m_0=1/5 and m_0=5. Lemma 3.3 says that the entries in the table's two columns are inverses, since the values of the m0s are inverses. The normal lines monotonically approach the lines $y = \pm x$, as shown by the arrows in their graphs in Figures 4 and 5.

Table 1: The slopes of the first eight iterated normal lines for p = 2.5		
	m0=1/5	**m0=5**
m1	−2.9240	−0.3120
m2	0.4890	2.0448
m3	−1.6111	−0.6207
m4	0.7277	1.3743
m5	−1.2361	−0.8090
m6	0.8682	1.1518
m7	−1.1000	−0.9101
m8	0.9391	1.0648

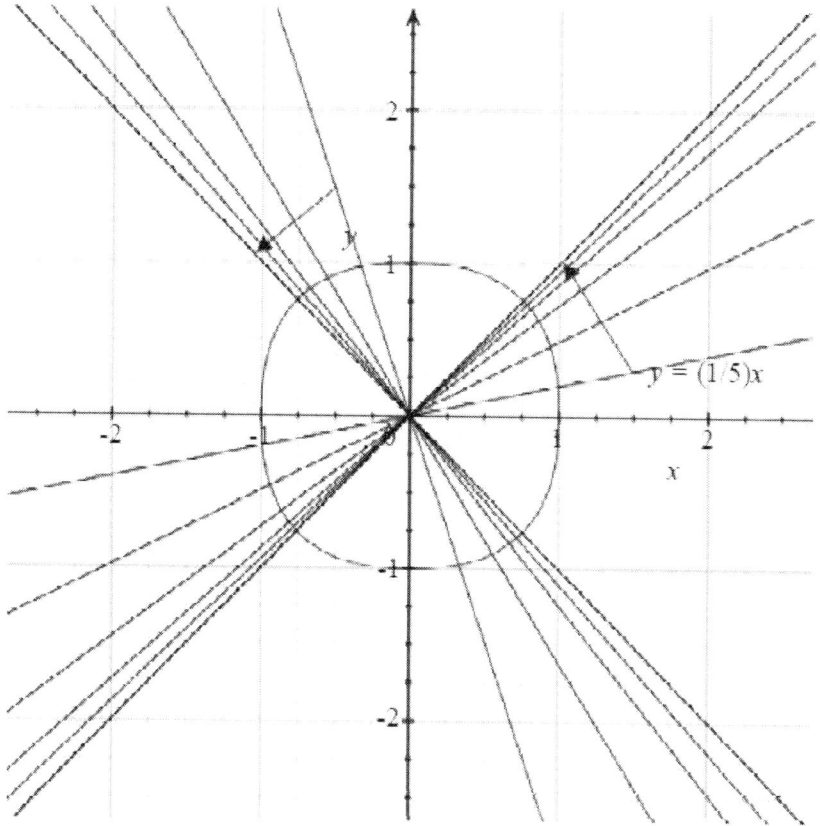

Figure 4: The lines $y=m_n x$ or the values in the first column of Table 1 for p=2.5 and $m_0=1/5$.

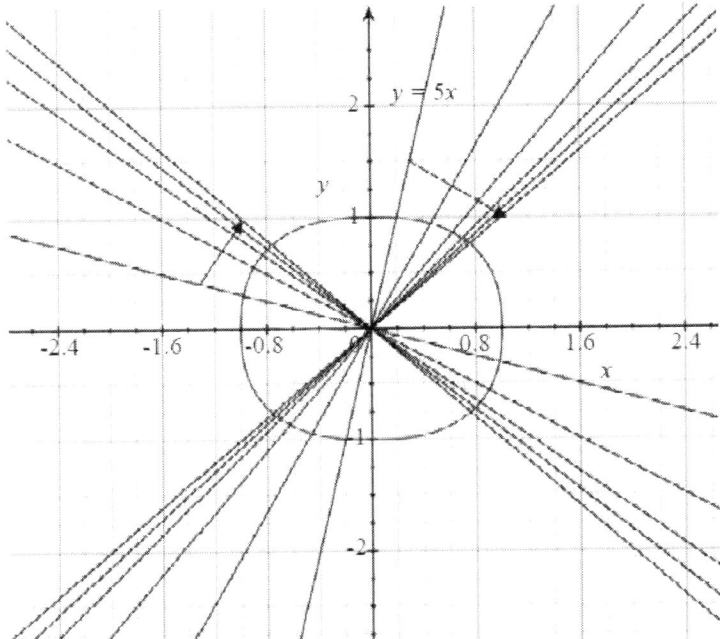

Figure 5: The lines $y=m_n x$ or the values in the second column of Table 1 for $p=2.5$ and $m_0=1/5$.

Theorem 3.5: Consider l_p geometry with $1<p<2$. For the initial line $y=m_0 x$ with $0<m_0<1$, the subsequence of $\{m_n\}$ for even n has values $0<m_n<1$ and monotonically approaches 0, and the subsequence for odd n has values $m_n<-1$ and monotonically approaches $-\infty$. For the initial line $y=m_0 x$ with $m_0>1$, the subsequence of $\{m_n\}$ for even n has values $m_n>1$ and monotonically approaches ∞, and the subsequence for odd n has values $m_n>-1$ and monotonically approaches 0.

Proof: The proof is the same as the proof of Theorem 3.4 with the small necessary changes being made.

As an example of Theorem 3.5, Table 2 contains the slopes of the first eight iterated normal lines for $p=5/3$ with $m_0=4/5$ and $m_0=5/4$. Lemma 3.3 says that the entries in the two columns are inverses, since the values of the m_0s are inverses. The normal lines monotonically approach the axes, as shown by the arrows in their graphs in Figure 6. The clockwise arrows are for $m_0=4/5$, and the counterclockwise arrows are for $m_0=5/4$.

The l_p geometries have unit circles that are symmetric about the lines $x=0$, $y=0$ and $y=-x$. Theorems 3.4 and 3.5 show that these directions are like attractors or else isolated pairs when iteratively taking normal lines. Taking $x=0$ or $y=0$ as the initial line gives a cycle of normal lines of period 2 between $x=0$ and $y=0$. Taking $y=x$ or $y=-x$ as the initial line gives a cycle of normal lines of period 2 between $y=x$ and $y=-x$.

Table 2: The slopes of the first eight iterated normal lines for $p=5/3$

	m0=4/5	m0=5/4
m1	−1.3975	−0.7155
m2	0.6053	1.6521
m3	−2.1236	−0.4709
m4	0.3231	3.0946
m5	−5.4439	−0.1837
m6	0.0787	12.702
m7	−45.269	−0.0221
m8	0.0033	304.58

A GEOMETRY WITH REFLEXIVE NORMALITY

Although our focus is on l_p geometries with $p>1$, portions of the unit circles (1) for different values of p can be joined to obtain interesting geometries. Theorem 4.1 shows how to make normality reflexive for all lines, that is, the normal to the normal of a line is the initial line. Reflexivity is sometimes called symmetry.

Theorem 4.1: Given the portion of the l_p unit circle that is in the first and third quadrants, the only way to complete a unit circle in the second and fourth quadrants for a Minkowski geometry with reflexive normality is with the portions of the l_q unit circle in the second and fourth quadrants for $1/p+1/q=1$.

Proof: Since Minkowski unit circles are symmetric about their centers, we can reference only the first and second quadrants. Take the center to be the origin, and construct all normal lines at the origin. In the

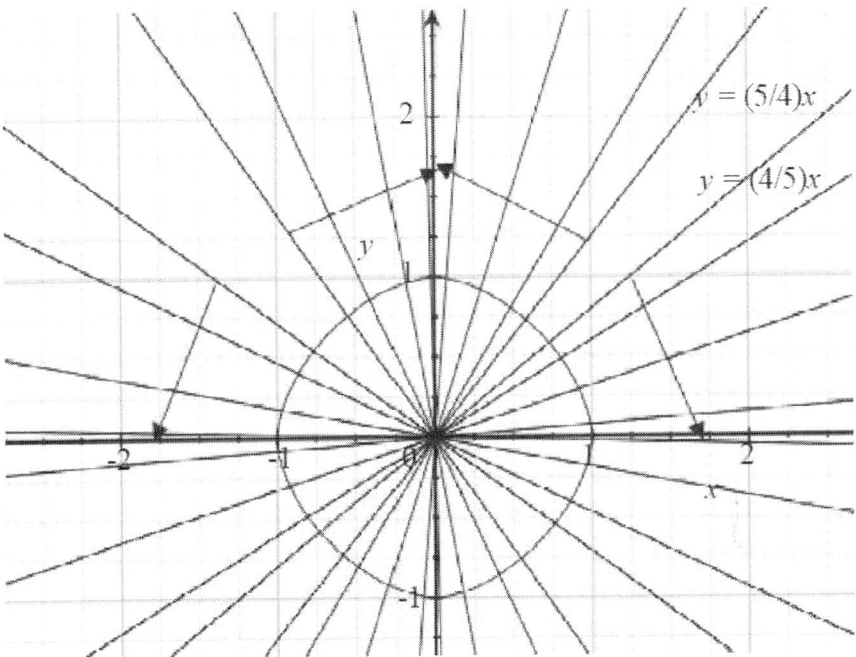

Figure 6: The lines $y=m_n x$ for the values in Table 2 for $p=5/3$ with $m_0=4/5$ and $m_0=5/4$.

first quad-rant, the unit circle is $x^p+y^p=1$. In the second quad-rant, the unit circle is $y=g(x)$. The original line L_1 is $y=tx$, $t>0$, which intersects $x^p+y^p=1$ at the point $p_1(x_1, y_1)$. The construction is illustrated in Figure 7 for $p = 4$. For reflexivity, demand that the slope of line L_1 equals the slope of the tangent line L_3 at the point $p_2(x_2, y_2)$. With $y_2=g(x_2)$, and demand that the slope of the line L_2 tangent to $x^p+y^p=1$ at $p_1(x_1, y_1)$ equals the slope y_2/x_2 of the line L_4, which is to be orthogonal to line L_1. The goal is to find the function $g(x)$. The slope of L_2 is found by taking the derivative of $x^p+y^p=1$ to obtain

$$px^{p-1} + py^{p-1}\, dy/dx = 0.$$

Then

$$dy/dx = -\left(x/y\right)^{|p-1|} = -\left\{x/(tx)\right\}^{p-1} = -1/t^{p-1} = -t^{1-p}.$$

Equating the slopes of the lines L_1 and L_3 gives

$$t = dy/dx(x_2)$$

12)

with y=g(x). Equating the slopes of lines L_2 and L_4 gives

$$-t^{1-p} = y_2/x_2 .$$

(13)

Solving (13) for t gives

$$t = \left(-y_2/x_2\right)^{1/(1-p)} .$$

(14)

Equating the expressions for t in (12) and (14) and dropping the subscript 2 give the differential equation

$$dy/dx = \left(-y/x\right)^{1/(1-p)} \quad \text{Or} \quad y^{1/(p-1)} \, dy/dx = \left(-x\right)^{1/(p-1)} ,$$

Whose unique solution is

Since y(0)=g(0)=1, C=1. Designating p/(p-1) by q gives 1/p+1/q=1 and

$$\left|x\right|^q + \left|y\right|^q = 1$$

For y=g(x) in the second and fourth quadrants.

The unit circles $|x|^4 + |y|^4 = 1$ and $|x|^{4/3} + |y|^{4/3} = 1$ are dual, since $1/(4)+1/(4/3)=1$. Dual unit circles and dual spaces are central to Minkowski geometry [1, 3, 7].

Schäffer's theorem says that dual unit circles have the same circumferences, when the circumferences are mea- sured with their own distance functions [1: pp. 111-118, 7: p. 153, 8, 9]. Because of the symmetry of the unit circle in Theorem 4.1, it has the same circumference as the dual I_p and I_q unit circles whose arcs compose it.

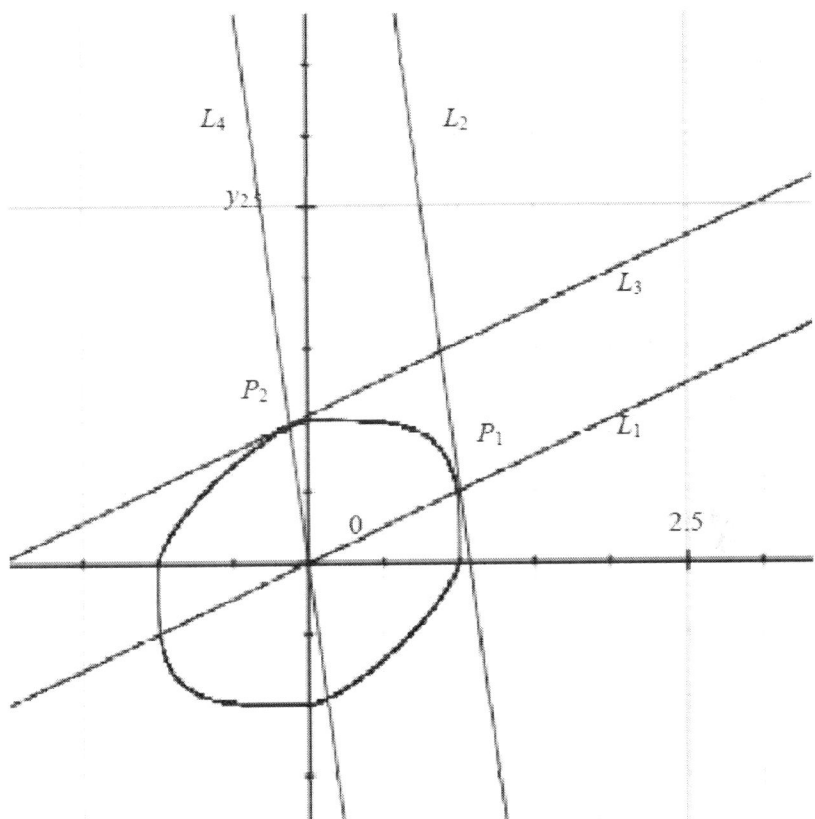

Figure 7: The unit circle in this Minkowski geometry is $|x|^4 + |y|^4 = 1$ in quadrants 1 and 3 and $|x|^{4/3} + |y|^{4/3} = 1$ in quadrants 2 and 4. Lines L_1 and L_3 are parallel, as are lines L_2 and L_4. In this geometry, normality is reflexive, that is, L_4 is normal to L_1 and L_1 is normal to L_4 for any choice of L1.

Radon curves are equivalently defined as either unit circles for which normality is reflexive for all lines or unit circles that have arcs of dual circles in alternating quadrants as in Theorem 4.1's example [1: p. 128, 3: pp. 233-234, 7: pp. 143-145, 10].

REFERENCES

1. C. Thompson, "Minkowski Geometry," Cambridge University Press, Cambridge, 1996. http://dx.doi.org/10.1017/CBO9781107325845.
2. E. F. Krause, "Taxicab Geometry," Dover Publications, New York, 1986.
3. R. V. Benson, "Euclidean Geometry and Convexity," McGraw-Hill, New York, 1966.
4. V. Dekster, "An Angle in Minkowski Space," Journal of Geometry, Vol. 80, No. 1, 2004, pp. 31-47.
5. G. J. Ruane and G. A. Swartzlander Jr., "Optical Vortex Coronagraphy with an Elliptical Aperture," Applied Optics, Vol. 52, No. 2, 2013, pp. 171-176. http://dx.doi.org/10.1364/AO.52.000171
6. M. Gardner, "Mathematical Carnival," Alfred A. Knopf, New York, 1975.
7. J. C. álvarez Paiva and A. Thompson, "On the Perimeter and Area of the Unit Disc," The American Mathematical Monthly, Vol. 112, No. 2, 2005, pp. 141-154. http://dx.doi.org/10.2307/30037412
8. J. J. Schaffer, "The Self-Circumferences of Polar Convex Disks," Archiv de Mathematik, Vol. 24, 1973, pp. 87-90. http://dx.doi.org/10.1007/BF01228179
9. J. B. Keller and R. Vakil, "pp, the Value of p in lp," The American Mathematical Monthly, Vol. 116, No. 10, 2009, pp. 931-935. http://dx.doi.org/10.4169/000298909X477069
10. J. Radon, "über eine Besondere Art Ebener Konvexer Kurven," Berichte der Sachsische Akademie der Wissenschaften zu Leipzig, Vol. 68, 1916, pp. 131-134.

CITATION

J. Fitzhugh and D. Farnsworth, "The Behavior of Normality when Iteratively Finding the Normal to a Line in an l_p Geometry," Advances in Pure Mathematics, Vol. 3 No. 8, 2013, pp. 647-652. doi: 10.4236/apm.2013.38086.

Algebra: Screw Blade Algorithm

Dimin Wu and Zhengzhi Wang

Institute of Automation, National University of
Defense Technology, Changsha, China.

2

ABSTRACT

The rigid body motion can be represented by a motor in geometric algebra, and the motor can be rewritten as a trinometric function of the screw blade. In this paper, a screw blade strapdown inertial navigation system (SDINS) algorithm is developed. The trigonometric function form of the motor is derived and utilized to deduce the Bortz equation of the screw blade. The screw blade SDINS algorithm is proposed by using the procedure similar to that of the conventional rotation vector attitude updating algorithm. The superiority of the screw blade algorithm over the conventional ones in precision is analyzed. Simulation results reveal that the screw blade algorithm is more suitable for the high-precision SDINS than the conventional ones.

INTRODUCTION

The strapdown inertial navigation system (SDINS) algorithm comprises attitude updating, velocity updating, and position updating, among which attitude updating usually utilizes the rotation vector algorithm [1-7]. The procedure of the rotation vector algorithm is: first, divide the attitude updating time interval equally into several segments and calculate the angular increment of each segment respectively; then, compute the total angular increment during the entire updating interval; finally, figure out the direction cosine matrix/quaternion of the updating time interval from the total angular increment. The theoretical foundation of the rotation vector algorithm is the Euler theorem, i.e., multiple rotations about axes passing through the origin is equivalent to a single rotation about some axis passing through the origin. Recently, a translation vector method, which is similar to the rotation vector algorithm, is proposed for the velocity/position updating [8].

According to the Chasles theorem, the rigid body motion can be represented by a rotation about an axis and a translation along that axis, and on this basis a SDINS algorithm based on dual quaternion (DQ) is presented [9]. The rotation vector and the translation vector are combined to construct the screw vector, which is utilized to calculate the DQs representing the vehicle's motions. All navigation parameters can be extracted from these DQs. The velocity/position updating algorithms in both [8,9] are related to the rotation vector algorithm which has been commonly used in modern SDINS algorithms. However, the attitude updating is combined with the velocity and position updating in [9], which is different from in [8] or other references, where the attitude updating is generally manipulated separately from the velocity and position updating. The benefit of the DQ-based SDINS algorithm is that the structure of the navigation algorithm can be simplified, and the precision can be promoted.

Geometric algebra (GA), as well as DQ, is a unitary representational tool for the rigid body motion, but it is more general than the latter. Rotations and translations of any dimension can be depicted in GA.

Motivated by [9], recently we have proposed another SDINS model using GA [10]. The navigation equations were recast into three alike motor kinematic equations. As a test of the GA-based SDINS model, the fourth-order Runge-Kutta (RK) method was chosen to solve the motor kinematic equations. The results revealed that the precision of the RK-based SDINS algorithm was quite high. However, the RK method is computationally costly, and it requires a good smoothness of the integral function. To meet the real-time requirement and the unavoidable measurement error in the SDINS, this paper is about to develop a

GA-based screw algorithm that is similar to the conventional rotation vector attitude updating algorithm.

The Contents of this paper are organized as follows. Section 2 reviews the GA SDINS model presented in [10]. Section 3 derives the trigonometric function form of the motor and the Bortz equation of the screw blade. A screw blade algorithm is developed in Section 4, and error analyses of the new algorithm and the conventional ones are presented in Section 5. In Section 6, a variety of simulations are carried out to testify the screw blade algorithm. Finally, concluding remarks are provided in Section 7.

GEOMETRIC ALGEBRA STRAPDOWN INERTIAL NAVIGATION SYSTEM MODEL

This section reviews the GA SDINS model, for detailed mathematical background please refer to [10] and references therein.

The purpose of inertial navigation is to provide navigation parameters through the integration of the angular rate and the total acceleration. The total acceleration consists of the specific force acceleration and the gravitational acceleration, which can be integrated into the thrust velocity and the gravitational velocity, respectively. Define the thrust velocity frame with its axes aligned with the body frames. The difference between the thrust velocity frame and the body frame is that the vector from the origin of the inertial frame to that of the thrust velocity frame is the thrust velocity, rather than the ordinary translation (See Figure 1). Similarly, we can define the gravitational velocity frame and the position frame. The axes of the gravitational velocity frame are aligned with those of the Earth-fixed frame. And the vector from the origin of the inertial frame to that of the gravitational velocity frame is the gravitational velocity. The position frame is attached to the centroid of the vehicle, and its axes are aligned with the Earth-fixed frames.

The kinematic equation of the motor that transforms the inertial frame I to the thrust velocity frame T can be formulated as

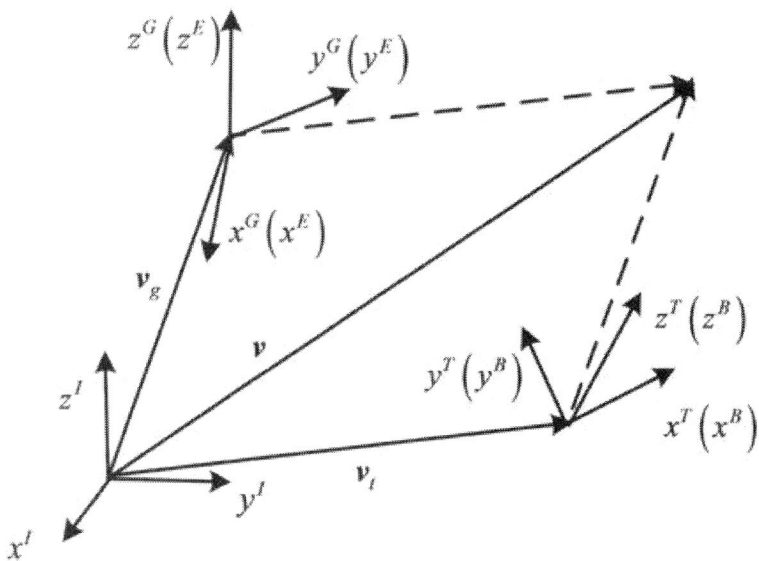

Figure 1: Velocity frames.

$$\dot{M}_{IT} = M_{IT}\Omega^{T}_{IT},$$ (1)

Where

$$\Omega^{T}_{IT} = \omega^{T*}_{IT} - \left(\tilde{R}_{IT}\upsilon^{I}_{t}R_{IT}\right)\infty = \omega^{B*}_{IB} - s^{B}\infty,$$ (2)

in which ω^{T}_{IT} and ω^{B}_{IB} denote the angular rates of the thrust velocity frame T and the body frame B with respect to the inertial frame I, respectively; υ^{I}_{t} denotes the thrust velocity; R_{IT} represents the rotor that transforms frame I to frame T; s^{B} is the specific force acceleration; * is the dual operator; $\tilde{}$ is the reverse operator.

Similarly, the kinematic equation of the motor that transforms the inertial frame I to the gravitational velocity frame G can be formulated as

$$2\dot{M}_{IG} = M_{IG}\Omega^{G}_{IG},$$ (3)

Where

$$\Omega_{IG}^{G} = \omega_{IG}^{G*} - \left(\tilde{R}_{IG} \upsilon_{g}^{I} R_{IG} \right) \infty = \omega_{IE}^{E*} - g^{E} \infty ,$$

(4)

in which ω_{IG}^{G} and ω_{IE}^{E} denote the angular rates of the gravitational velocity frame G and the Earth-fixed frame E with respect to the inertial frame I, respectively; υ_{g}^{I} denotes the gravitational velocity; R_{IG} represents the rotor that transforms frame I to frame G; g^{E} is the gravitational acceleration.

The kinematic equation of the motor that transforms the inertial frame I to the position frame p can be formulated as

$$2\dot{M}_{IP} = M_{IP}\Omega_{IP}^{P} ,$$

(5)

Where

$$\Omega_{IP}^{P} = \omega_{IP}^{P*} - \left(\tilde{R}_{IP} \dot{r}^{I} R_{IP} \right) \infty = \omega_{IE}^{E*} - \left(\tilde{R}_{IG} \left(v_{I}^{I} + v_{g}^{I} \right) R_{IG} \right) \infty$$

(6)

in which ω_{IP}^{P} denotes the angular rate of frame P with respect to frame I; r^{I} denotes the position vector from the origin of frame I to that of frame p; R_{IP} represents the rotor that transforms frame I to frame P. The mechanism of the GA SDINS model is shown in Figure 2.

SCREW BLADE BORTZ EQUATION

As shown in the last section, the kinematic equation of the motor has the unitary form

$$\dot{M} = \frac{1}{2}M\Omega ,$$

(7)

where $\Omega = \omega * - \upsilon\infty$ denotes the screw velocity which consists of the angular rate ω and the velocity υ. It can be solved by traditional numerical integral algorithms. However, in order to meet the real-time requirements of the SDINS, a high-efficiency algorithm, which is simi-

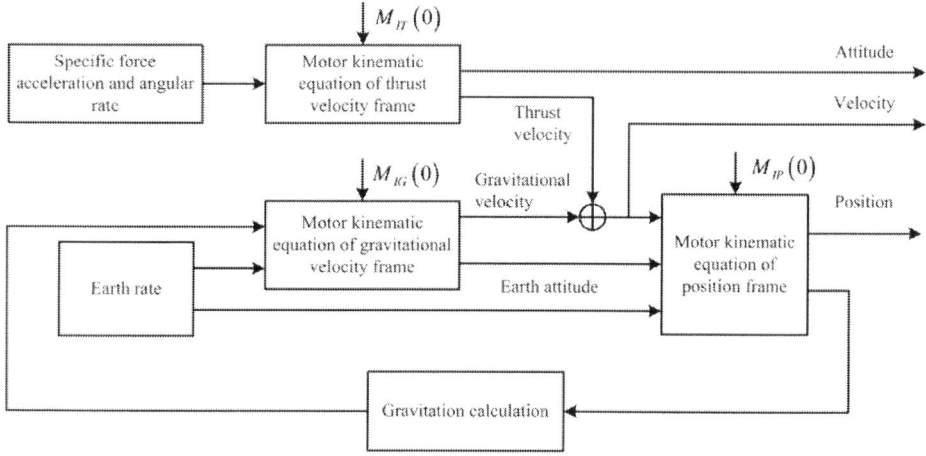

Figure 2: Mechanism of the GA SDINS model.

lar to the rotation vector algorithm that has been widely used in the modern SDINS algorithm, needs to be developed. The rotation vector algorithm is based on the Bortz equation; therefore, we will build the GA-formed Bortz equation first.

A general rigid body motion can be constructed by a rotation in a plane around the origin, followed by a translation. As shown in Figure 3, the motor which represents the general rigid body motion, can be decomposed into the screw parameters as [11]

$$M = T_t R_{N\theta} = T_w R_{T_v[N\theta]} = e^{-w\infty/2 - T_v[N\theta]/2},$$

(8)

where $T_{(\cdot)} = e^{-(\cdot)\infty/2}$ and $R_{(\cdot)} = e^{-(\cdot)/2}$ represent the translation versor and the rotation versor (or rotor), respectively; $T\left[(\cdot)\right] = T(\cdot)^{T-1}$ represents the versor product;

t denotes the translation; N denotes the plane in which the rotation is resident, and $N^2 = -1$; θ is the rotation angle; $w = (t \wedge N)/N$ is the translation along the rotation (or screw) axis; $v = \left(1 - R_{N\theta}^2\right)^{-1}(t \cdot N)/N$ is the translation of the rotation axis.

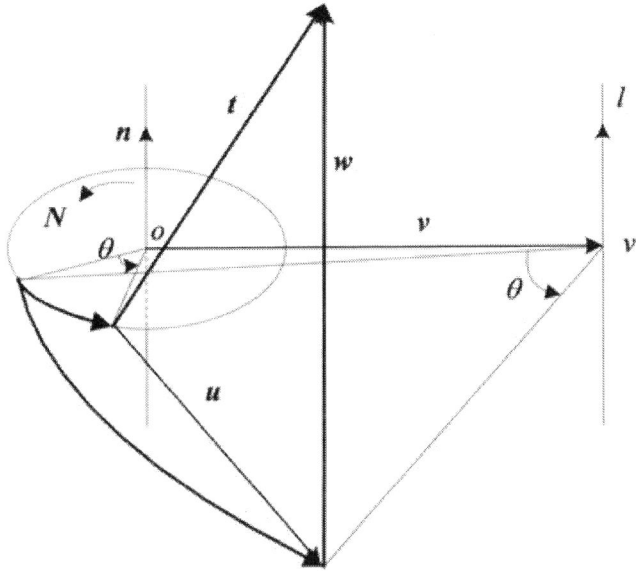

Figure 3: Screw decomposition of the motor.

Define the screw blade to be $\Theta = -w\infty - T_v[N\theta]$, we have

$$\Theta = -w\infty - T_v[N\theta] = -w\infty - T_v N\theta T_v^{-1} = -N\theta$$
$$+ (v \cdot N\theta - w)\infty = (\theta - dI\infty)(n^* - v \cdot n^*\infty),$$

(9)

Where I is the pseudoscalar; $n=N^*=N \cdot I^{-1}$ denotes the direction of the screw axis; $d=w \cdot n=wn$ denotes the magnitude of w. It can be seen that $n^*-v \cdot n^*\infty$ is the right dual representation of the screw axis lz. Define $\Theta = \theta - dI\infty$, it can be shown that

$$\Theta = \|\Theta\| = (\Theta\tilde{\Theta})^{1/2} = (-\Theta^2)^{1/2}.$$

(10)

Therefore, Equation (9) can be simplified as $\Theta = \Theta I$, or, $l = \Theta / \Theta$. Considering that ∞ to the power of two or more equals zero, the motor can be rewritten as [12]

$$M = e^{\Theta/2} = e^{\Theta/2} = \cos\frac{\Theta}{2} + I\sin\frac{\Theta}{2}$$

$$= \cos\frac{\Theta}{2} + \frac{\Theta}{\Theta}\sin\frac{\Theta}{2},$$

(11)

Where

$$\cos\Theta = 1 - \frac{\Theta^2}{2!} + \frac{\Theta^4}{4!} - \cdots = 1 - \frac{\theta^2 - 2\theta dI\infty}{2!}$$

$$+ \frac{\theta^4 - 4\theta^3 dI\infty}{4!} - \cdots = \cos\theta + dI\infty\sin\theta,$$

(12)

$$\sin\Theta = \Theta - \frac{\Theta^3}{3!} + \frac{\Theta^5}{5!} - \cdots = \sin\theta - dI\infty\cos\theta.$$

(13)

It is seen that the motor M is determined by the screw blade Θ; therefore, M can be obtained if Θ is known. Differentiating Equation (11) with respect to time gives

$$\dot{M} = -\frac{\dot{\Theta}}{2}\sin\frac{\Theta}{2} + \frac{\dot{\Theta}}{\Theta}\sin\frac{\Theta}{2} - \frac{\Theta\dot{\Theta}}{2\Theta^2}\left(2\sin\frac{\Theta}{2} - \Theta\cos\frac{\Theta}{2}\right).$$

(14)

Substituting Equation (11) into the right side of Equation (7), yields

$$\frac{1}{2}M\Omega = \frac{1}{2}\Omega\cos\frac{\Theta}{2} + \frac{\Theta\Omega + \Omega\Theta + \Theta\Omega - \Omega\Theta}{4\Theta}\sin\frac{\Theta}{2}$$

$$= \frac{\Theta\Omega + \Omega\Theta}{4\Theta}\sin\frac{\Theta}{2} + \frac{1}{2}\Omega\cos\frac{\Theta}{2} + \frac{\Theta\times\Omega}{2\Theta}\sin\frac{\Theta}{2},$$

(15)

Where

$$\Theta\times\Omega = \frac{\Theta\Omega - \Omega\Theta}{2}$$

Is the cross product of two blades [11-13].

With Equations (7), (14), and (15), considering that terms of the same grade on one side of the equation should equal to those on the other side, it gives

$$\dot{\Theta} = -\frac{\Theta\Omega + \Omega\Theta}{2\Theta},$$

(16)

$$\frac{\dot{\Theta}}{\Theta}\sin\frac{\Theta}{2} - \frac{\Theta\dot{\Theta}}{2\Theta^2}\left(2\sin\frac{\Theta}{2} - \Theta\cos\frac{\Theta}{2}\right) = \frac{1}{2}\Omega\cos\frac{\Theta}{2}$$
$$+ \frac{\Theta\times\Omega}{2\Theta}\sin\frac{\Theta}{2}.$$

(17)

Substituting Equations (10) and (16) into Equation (17) gives

$$\dot{\Theta} = \Omega + \frac{1}{2}\Theta\times\Omega + \frac{\Theta(\Theta\times\Omega)}{\Theta^2}\left(1 - \frac{\Theta}{2}\cot\frac{\Theta}{2}\right).$$

(18)

In fact, when t=0 and $\upsilon = 0$, it follows that

$$d = 0, \quad l = n^*, \quad \Theta = \theta, \quad \Omega = \omega^*.$$

(19)

Define $\theta = \theta n$, we have

$$\Theta = \Theta l = \theta n^* = \theta^*.$$

(20)

Introducing Equations (19) and (20) into Equation (18), and multiplying the resultant equation with the pseudoscalar I, Equation (18) can be simplified as

$$\dot{\theta} = \omega + \frac{1}{2}\theta\times\omega + \frac{\theta\times(\theta\times\omega)}{\theta^2}\left(1 - \frac{\theta}{2}\cot\frac{\theta}{2}\right),$$

(21)

Which is the Bortz equation of the rotation vector [1]. Therefore, Equation (18) can be considered as an extension of the Bortz equation, i.e., the Bortz equation of the screw blade.

SCREW BLADE ALGORITHM

Since the updating time interval is very short, Θ and $\dot{\Theta}$ are small. Neglecting the second and higher orders of Θ and $\dot{\Theta}$, Equation (18) can be simplified as

$$\dot{\Theta} = \Omega + \frac{1}{2}\Theta \times \Omega .$$

$$(22)$$

Therefore, the screw blade increment of a time interval h can be computed as

$$\Theta = \int_0^h \Omega(t)\,dt + \Delta\Theta_c .$$

$$(23)$$

In which $\Delta\Theta_c$ is the integral of the cross product, and it can be approximated as a noncommutative rate vector of the rotation vector [2,6]:

$$\Delta\Theta_c = \frac{1}{2}\int_0^h \Delta\Theta(t) \times \Omega(t)\,dt ,$$

$$(24)$$

Where

$$\Delta\Theta(t) = \int_0^t \Omega(\tau)\,d\tau$$

On the other hand, the solution of Equation (22) in Taylor series is given by

$$\Theta(h) = \Theta(0) + h\dot{\Theta}(0) + \cdots .$$

$$(25)$$

Assume that the screw velocity can be represented by a third-order polynomial

$$\Phi(t) = \int_0^t \Omega(\tau)\,d\tau = At + Bt^2, \quad 0 \leq t \leq h,$$

$$(26)$$

where A and B are constant blades. The 2-sample screw blade formula can be acquired by the following procedure:

1. Compute the differentiations of $\Theta\,(0)$, and substitute them into Equation (25);
2. Divide the time interval evenly into 2 sub-intervals, and calculate the screw blade increment of each subinterval by integral of the screw velocity;
3. Compute the coefficients of the polynomial using the screw blade increments of the sub-intervals, and then substitute them into Equation (25).

It is found that

$$\Delta\Theta_c = \frac{2}{3}\Delta\Theta_1 \times \Delta\Theta_2,$$

$$(27)$$

Where $\Delta\Theta_1$ and $\Delta\Theta_2$ denote the screw blade increments of the sub-intervals. Other multiple samples screw blade formulae can be set up similarly. A general Nsample screw blade algorithm can be formulated as

$$\Delta\Theta_c = \sum_{i=1}^{N} \sum_{j=i+1}^{N} K_{ij}\Delta\Theta_i \times \Delta\Theta_j,$$

$$(28)$$

Where the constant coefficients k_{ij} are the same as those in the rotation vector algorithms.

The motor of the updating interval can thus be calculated as

$$\Delta M = \cos\frac{\Theta}{2} + \frac{\Theta}{\Theta}\sin\frac{\Theta}{2},$$

$$(29)$$

Where $\Theta = \|\Theta\|$. The total motor can be updated as

$$M_{n+1} = M_n \Delta M,$$

$$(30)$$

In which M_n and M_{n+1} represent the previous and current motors, respectively.

ALGORITHM ERROR ANALYSIS

Since Θ is the integral of Ω, we can assume $\Theta = \theta^* - s\infty$, where $\Theta = \int \omega dt$, $s = \int \upsilon dt$. It follows that

$$\Theta = \left(-\Theta^2\right)^{1/2} = \left(\theta^2 - 2(s \cdot \theta) I\infty\right)^{1/2} = \theta - dI\infty, \tag{31}$$

Where $\theta = \left(\theta^2\right)^{1/2}$ the magnitude of the rotation angleand is $d = s \cdot \theta / \theta$ is the magnitude of the translation along the screw axis. Therefore, Equation (11) can be expanded as

$$M = \cos\frac{\Theta}{2} + \frac{\Theta}{\Theta}\sin\frac{\Theta}{2} = \cos\frac{\theta}{2} + \frac{d}{2} I\infty \sin\frac{\theta}{2}$$
$$+ \frac{\theta^* - s\infty}{\theta - dI\infty}\left(\sin\frac{\theta}{2} - \frac{d}{2} I\infty \cos\frac{\theta}{2}\right) = \cos\frac{\theta}{2} + \frac{\theta^*}{\theta}\sin\frac{\theta}{2}$$
$$- \left(-dI\sin\frac{\theta}{2} + \frac{2s}{\theta}\sin\frac{\theta}{2} + \frac{d\theta}{\theta}\cos\frac{\theta}{2} - \frac{2d\theta}{\theta^2}\sin\frac{\theta}{2}\right)\frac{\infty}{2}. \tag{32}$$

On the other hand, the motor increment of the interval can be formulated as [11,12]

$$\Delta M = R - tR\infty/2. \tag{33}$$

Therefore, the translation increment of the updating interval can be calculated as

$$t = \left(-dI\sin\frac{\theta}{2} + \frac{2s}{\theta}\sin\frac{\theta}{2} + \frac{d\theta}{\theta}\cos\frac{\theta}{2} - \frac{2d\theta}{\theta^2}\sin\frac{\theta}{2}\right)$$
$$\times \left(\cos\frac{\theta}{2} + \frac{\theta^*}{\theta}\sin\frac{\theta}{2}\right) = -\frac{d}{2} I\sin\theta + s\frac{\sin\theta}{\theta}$$
$$+ \frac{d\theta}{2\theta}(1 + \cos\theta) - \frac{d\theta}{\theta^2}\sin\theta + \frac{d\theta}{2\theta}(1 - \cos\theta)$$
$$- \frac{s\theta^*}{\theta^2}(1 - \cos\theta) + \frac{d}{2} I\sin\theta - \frac{d}{\theta} I(1 - \cos\theta)$$

$$= \frac{\mathrm{d}\theta}{\theta}\left(1 - \frac{\sin\theta}{\theta}\right) + s\frac{\sin\theta}{\theta} + \frac{(\theta \wedge s)^*}{\theta^2}(1 - \cos\theta)$$

(34)

The relationship

$$s\theta^* = s\theta(-I) = -(s \cdot \theta)I - (s \wedge \theta)I = -\mathrm{d}\theta I - (\theta \wedge s)^*$$

Is utilized in the above derivation.

Approximating the trigonometric functions by second order Taylor expansion, Equation (34) can be reduced to be

$$t = s + \frac{1}{2}(\theta \wedge s)^* = s + \frac{1}{2}\theta \wedge s.$$

(35)

With Equations (22) and (28), the screw blade increment of the updating interval can be computed as

$$\begin{aligned}
\Theta &= \sum_{i=1}^{N}\Theta_i + \sum_{i=1}^{N}\sum_{j=i+1}^{N} K_{ij}\Theta_i \times \Theta_j \\
&= \sum_{i=1}^{N}\theta_i^* - \sum_{i=1}^{N}\sum_{j=i+1}^{N} K_{ij}\theta_i \wedge \theta_j \\
&\quad - \left(\sum_{i=1}^{N}s_i + \sum_{i=1}^{N}\sum_{j=i+1}^{N} K_{ij}\left(\theta_i \wedge s_j + s_i \wedge \theta_j\right)\right)^* \infty,
\end{aligned}$$

(36)

Where $\Theta_i = \theta_i^* - s_i\infty$ is the screw blade increment of the sub-interval. Substituting (36) into (35) gives

$$\begin{aligned}
t &= \sum_{i=1}^{N}s_i + \sum_{i=1}^{N}\sum_{j=i+1}^{N} K_{ij}\left(\theta_i \wedge s_j + s_i \wedge \theta_j\right)^* \\
&\quad + \frac{1}{2}\left(\sum_{i=1}^{N}\theta_i^* - \sum_{i=1}^{N}\sum_{j=i+1}^{N} K_{ij}\theta_i \wedge \theta_j\right)^{-*} \\
&\quad \times \left(\sum_{i=1}^{N}s_i + \sum_{i=1}^{N}\sum_{j=i+1}^{N} K_{ij}\left(\theta_i \wedge s_j + s_i \wedge \theta_j\right)^*\right)
\end{aligned}$$

$$= \sum_{i=1}^{N} s_i + \sum_{i=1}^{N} \sum_{j=i+1}^{N} K_{ij} \left(\theta_i \times s_j + s_i \times \theta_j \right)$$

$$+ \frac{1}{2} \left(\sum_{i=1}^{N} \theta_i + \sum_{i=1}^{N} \sum_{j=i+1}^{N} K_{ij} \theta_i \times \theta_j \right)$$

$$\times \left(\sum_{i=1}^{N} s_i + \sum_{i=1}^{N} \sum_{j=i+1}^{N} K_{ij} \left(\theta_i \times s_j + s_i \times \theta_j \right) \right).$$

(37)

Its second-order approximation is

$$t = \sum_{i=1}^{N} s_i + \sum_{i=1}^{N} \sum_{j=i+1}^{N} K_{ij} \left(\theta_i \times s_j + s_i \times \theta_j \right) + \frac{1}{2} \sum_{i=1}^{N} \theta_i \times \sum_{i=1}^{N} s_i$$

(38)

The last two terms in the above Equation are the sculling and rotation compensation terms in the conventional velocity integration [14,15]. Therefore, the error for the conventional algorithm includes: 1) approximation of the screw blade during the updating time interval in Equation (36); 2) truncation of trigonometric series to the second order in Equation (35); 3) approximation to second-order accuracy in Equation (38). On the other hand, the error for the screw blade algorithm comes from: 1) approximation of the screw blade in Equation (36); and 2) truncation of trigonometric functions in calculating Equation (34). Therefore, if the error in Equation (36) is small enough, the precision of the screw blade algorithm will be higher than that of the conventional ones. The error arisen from the approximation of the screw blade can be reduced by choosing high-precision inertial sensors to decrease measurement error, or by using big sample screw blade algorithms, or by shortening updating time intervals, etc.

NUMERICAL SIMULATION

To testify the performance of the screw blade algorithm presented in this paper, a variety of experiments are carried out.

The navigation parameters and the gyro/accelerometer outputs are produced by an ideal trace generator. The inputs of the generator are the ground speed rate and the Euler angle rate, and the inputs for each channel are the same, which are $20\sin(2\pi ft)$ (m/s²) for the former, and $2\pi f \cos(2\pi ft)$ (rad/s) for the latter. The frequencies are f = 0.01, 0.02, 0.04, 0.06, 0.08, 0.1, 0.2, 0.4, 0.6 and 0.8 (Hz). At the beginning, the vehicle stands still on the ground of the Earth at latitude 30° and longitude 110°, and the attitude angles are all 0. The runtime is set to be 300 (s).

The coordinate transformation between the Earth fixed frame coordinate and the geodetic coordinate is achieved by an iterative algorithm proposed in [16]. To increase the precision, the gravitational acceleration is computed using the information of real time position of the vehicle. Since the position is calculated only once per updating interval, a third-order algorithm [17] is utilized in computing the screw blades of the gravitational velocity frame motor and the position frame motor.

In this test, the 2-sample screw blade algorithm is chosen to solve the GA SDINS model. For the conventional algorithm, 2-sample formulae are applied in attitude/velocity updating [4,18], and the trapezoidal integration is utilized in position updating.

Here the measurement errors of the inertial sensors are simplified as a compound effect of the fixed bias and the Gaussian random noise. Four typical SDINS configurations used in the test are listed inTable 1. The maximum absolute errors (MAE) of both algorithms are illustrated in Figure 4 as a function of the varying input signal frequency. "CA" and "GA" stand for the conventional algorithm and the geometric algebra screw blade algorithm, respectively. The updating time interval is 0.02 (s).

It can be seen that:

1. The attitude errors of both algorithms are the same, but the velocity and position errors have been remarkably reduced due to the combination of the attitude updating and the velocity/position updating, which utilizes the highprecision attitude updating algorithm.

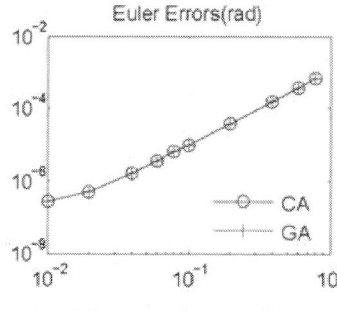
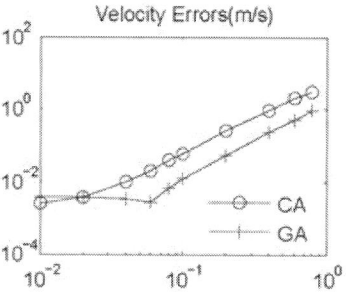
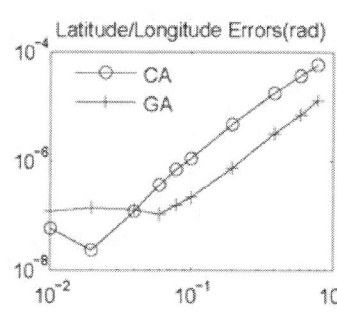
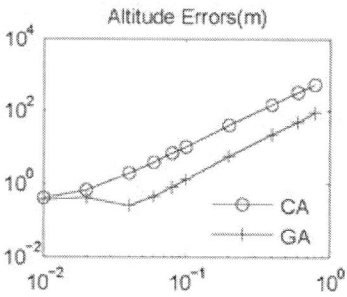

(a)

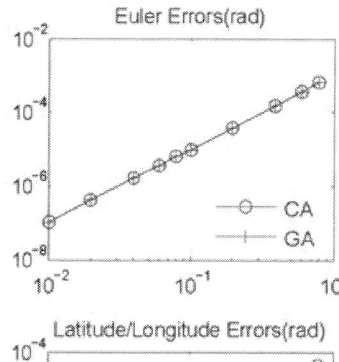
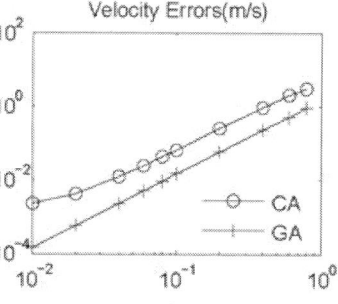
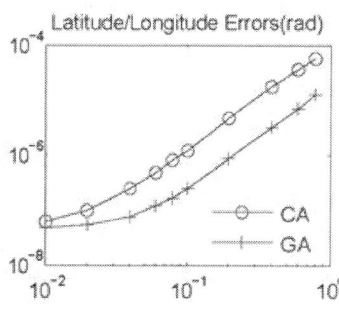
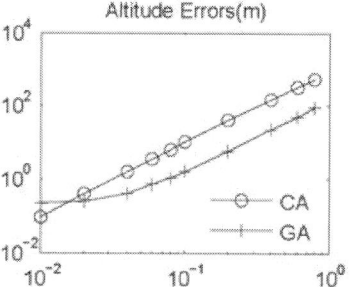

(b)

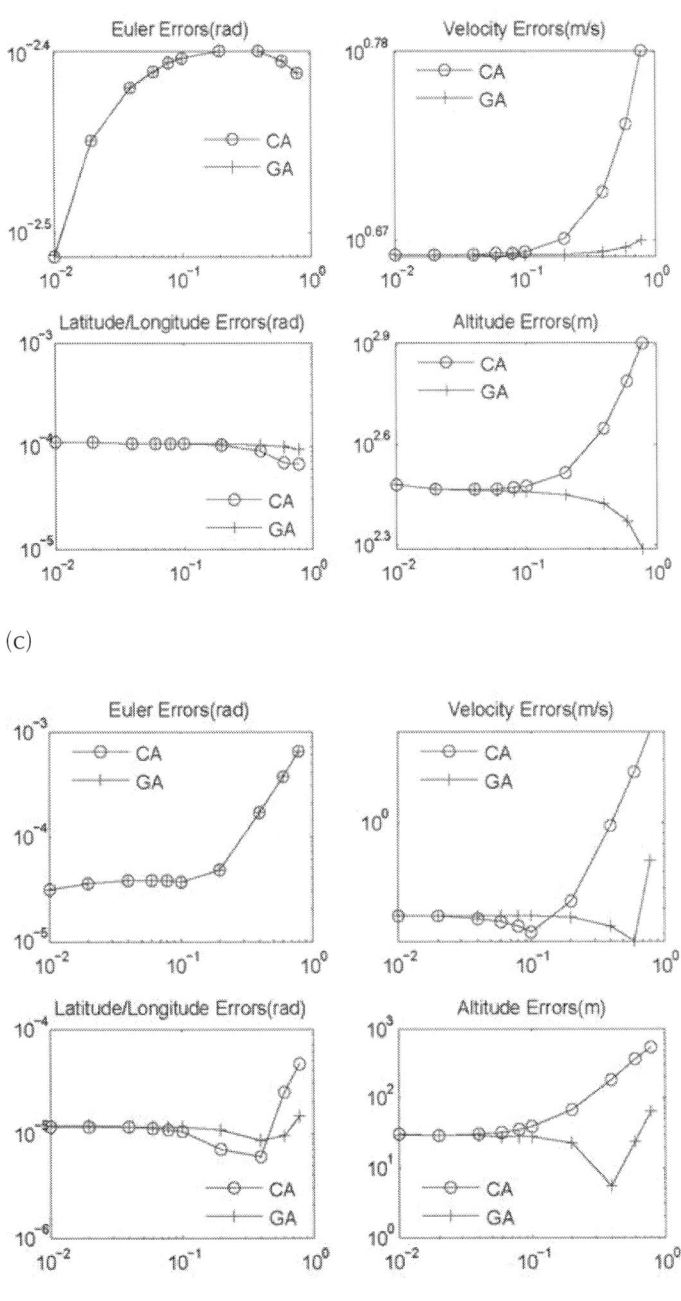

(c)

(d)

Figure 4: MAEs of algorithms with variant measurement errors as a function of varying input signal frequency. (a) Test 1; (b) Test 2; (c) Test 3; (d) Test 4.

2. When there is no measurement error (Figure 4(a)), or the mea-surement error is minor (Figure 4(b)), the superiority of the GA over the CA is quite obvious. However, as the measurement error grows, the superiority loses gradually, as shown in Figure 4(d), in which the MAEs of both algorithms are almost the same. This is because the measurement error becomes the main error source of the algorithms as it grows.

3. There are some turning frequency points around which the MAEs of the CA are smaller than those of the GA. However, in most cases, the precision of the GA is better than that of the CA.

Table 1: Inertial sensor configuration

		Gyro		Accelerometer	
	Fixed bias (/h)	Gaussian noise (/h1 σ)	Fixed bias(g)	Gaussian noise (g1 σ)	
Test 1	0	0	0	0	
Test 2	10^{-4}	10^{-5}	1	10-1	
Test 3	10^{-2}	10^{-3}	10^2	10	
Test 4	1	10^{-1}	10^3	10^2	

CONCLUSIONS

A screw blade algorithm is proposed to solve the GA SDINS model. At first, the trigonometric function form of the motor is derived by us-ing the screw decomposition. The Bortz equation of the screw blade is deduced in succession. And then, the screw blade SDINS algorithm is developed similar to the conventional rotation vector attitude updating algorithm. The errors of the screw blade algorithm and the conven-tional ones are analyzed as well.

The performances of the screw blade algorithm are testified by a va-riety of simulations. The results reveal that the screw blade algorithm is a better choice than the conventional ones when the measurement error is small, and it is more suitable for the SDINS with high-precision requirement.

ACKNOWLEDGEMENTS

This project is supported by the National Natural Science Foundation of China (60835005). We appreciate Dr. Danfeng Zhou for his constructive suggestions in scientific writing.

REFERENCES

1. J. E. Bortz, "A New Mathematical Formulation for Strapdown Inertial Navigation," IEEE Transactions on Aerospace and Electronic Systems, Vol. AES-7, No. 1, 1971, pp. 61-66.doi:10.1109/TAES.1971.310252
2. M. B. Ignagni, "Optimal Strapdown Attitude Integration Algorithms," Journal of Guidance, Vol. 13, No. 2, 1990, pp. 363-369. doi:10.2514/3.20558
3. M. B. Ignagni, "Efficient Class of Optimized Coning Compensation Algorithm," Journal of Guidance, Control, and Dynamics, Vol. 19, No. 2, 1996, pp. 424-429. doi:10.2514/3.21635
4. J. W. Jordan, "An Accurate Strapdown Direction Cosine Algorithm," NASA TN-D-5384, 1969.
5. J. G. Lee, Y. J. Yoon, J. G. Mark and D. A. Tazartes, "Extension of Strapdown Attitude Algorithm for High-Frequency Base Motion," Journal of Guidance, Vol. 13, No. 4, 1990, pp. 738-743. doi:10.2514/3.25393
6. R. B. Miller, "A New Strapdown Attitude Algorithm," Journal of Guidance, Vol. 6, No. 4, 1983, pp. 287-291. doi:10.2514/3.19831
7. P. G. Savage, "Strapdown Inertial Navigation Integration Algorithm Design Part 1: Attitude Algorithms," Journal of Guidance, Control, and Dynamics, Vol. 21, No. 1, 1998, pp. 19-28. doi:10.2514/2.4228
8. P. G. Savage, "A Unified Mathematical Framework for Strapdown Algorithm Design," Journal of Guidance, Control, and Dynamics, Vol. 29, No. 2, 2006, pp. 237-249.doi:10.2514/1.17112
9. Y. Wu, X. Hu, D. Hu, T. Li and J. Lian, "Strapdown Inertial Navigation System Algorithms Based on Dual Quaternion," IEEE Transactions on Aerospace and Electronic Systems, Vol. 41, No. 1, 2005, pp. 110-132. doi:10.1109/TAES.2005.1413751
10. D. Wu and Z. Wang, "Strapdown Inertial Navigation System Algorithm Based on Geometric Algebra," Advances in Applied Clifford Algebras, Vol. 22, No. 1, 2012, pp. 1- 17.
11. L. Dorst, D. Fontijne and S. Mann, "Geometric Algebra for Computer Science: An Object-Oriented Approach to Geometry," 2nd Edition, Morgan Kaufmann Publishers, San Francisco, 2007.
12. E. Bayro-Corrochano, "Geometric Computing for Wavelet Transforms, Robot Vision, Learning, Control and Action," Springer Verlag, London, 2010.

13. D. Hestenes, "New Foundations for Classical Mechanics," 2nd Edition, Kluwer Academic Publishers, Dordrecht, 1999.

14. M. B. Ignagni, "Duality of Optimal Strapdown Sculling and Coning Compensation Algorithms," Journal of the Institute of Navigation, Vol. 45, No. 2, 1998, pp. 85-95.

15. P. G. Savage, "Strapdown Inertial Navigation Integration Algorithm Design Part 2: Velocity and Position Algorithms," Journal of Guidance, Control, and Dynamics, Vol. 21, No. 2, 1998, pp. 208-221. doi:10.2514/2.4242

16. Y. Wu, P. Wang and X. Hu, "Algorithm of Earth-Centered Earth-Fixed Coordinates to Geodetic Coordinates," IEEE Transaction on Aerospace and Electronic Systems, Vol. 39, No. 4, 2003, pp. 1457-1461. doi:10.1109/TAES.2003.1261144

17. R. A. McKern, "A Study of Transformation Algorithms for Use in a Digital Computer," M.S. Thesis, MIT, Cambridge, 1968.

18. Y. Qin, "Inertial Navigation (in Chinese)," Science publisher, Beijing, 2006.

CITATION

D. Wu and Z. Wang, "Strapdown Navigation Using Geometric Algebra: Screw Blade Algorithm," Positioning, Vol. 3 No. 2, 2012, pp. 13-20. doi: 10.4236/pos.2012.32003.

On q-Deformed Calculus in Quantum Geometry

Olaniyi S. Maliki[1], Emmanuel I. Ugwu[2]

[1]Department of Industrial Mathematics and Applied Statistics, Ebonyi State University, Abakaliki, Nigeria
[2]Department of Industrial Physics, Ebonyi State University, Abakaliki, Nigeria

ABSTRACT

The relation between noncommutative (or quantum) geometry and the mathematics of spaces is in many ways similar to the relation between quantum physics and classical physics. One moves from the commutative algebra of functions on a space (or a commutative algebra of classical observable in classical physics) to a noncommutative algebra representing a noncommutative space (or a noncommutative algebra of quantum observables in quantum physics). The object of this paper is to study the basic rules governing q-calculus as compared with the classical Newton Leibnitz calculus.

INTRODUCTION

There exists an intimate relationship between classical geometry and physics, and to appreciate this we will consider the Einstein field equations (EFE) of special relativity written as:

$$G_{\mu v} = 8\pi T_{\mu v} \qquad\qquad \text{(EFE)}$$

Where $G_{\mu v}=R_{\mu v}-\dfrac{1}{2}g_{\mu v}R$ is the Einstein tensor, $R=g^{\mu v}R_{\mu v}$ is the Ricci tensor. $T_{\mu v}$ is the energy-momentum tensor.

For the moment we are only interested in three basic properties of the above equation.

1) The equation (EFE) is a tensor equation. This is necessarily so, since the principle of invariance under coordinate transformations must hold, in other words the equations of physics must look the same in any frame of reference.

2) We can interpret equation (EFE) more simply as

Tensor representing geometry of space = Tensor representing energy content of space

i.e. it is the presence of matter in space that distorts the neighbouring geometry. Most equations of mathematical physics can be interpreted similarly.

3) The solution to equation (EFE) is a geometrical object, namely a line element given by

$$ds^2 = g_{\mu\nu}dx^\mu dx^\nu$$

Where $g_{\mu\nu}$ is the metric tensor to be solved for in (EFE).

Quantum Geometry

Every geometry is associated with some kind of space. Quantum (or noncommutative) geometry [1][2] deals with quantum spaces, including the classical concept of space as a very special case. In classical geometry spaces are always regarded as collections of points equipped with the appropriate additional structure (as for example a topological structure given by the collection of open sets, or a smooth structure given by the atlas). In contrast to classical geometry, quantum spaces are not interpretable in this way. In general, quantum spaces have no points at all! They exhibit non-trivial quantum fluctuations' of geometry at all scales.

In generalizing classical geometry to the non-commutative level, there are two important conceptual steps:

1. Translation of geometry into a commutative algebra format;
2. Non-commutative generalizations.

Reformulating Basic Geometrical Concepts

It turns out that the geometrical structure on any given topological space X is always completely expressible in the language of some associated *-algebra [3] [4].

Points

Let X be a compact topological space, and let A = C(X) be the *-algebra of continuous complex-valued functions on X. Every element x∈ X naturally gives rise to a linear functional $\chi = \chi_x : A \to \mathbb{C}$ defined by

$$\chi(f) = f(x)$$

This map is multiplicative in the sense that

$$\chi(fg) = \chi(f)\chi(g), \qquad \forall f, g \in A$$

Furthermore, c is Hermitian in the sense that $\chi(f^*) = \chi(f)^*$ it is also non-zero, i.e. x is a character on A. Conversely, consider an arbitrary character $\chi : A \to \mathbb{C}$, then it can be shown that there exists a unique point x∈X such that $\chi = \chi_x$. In other words, we have a natural bijection between points of X and characters of A. It is important to note that this characterization of points also remains valid at the smooth level, in which X could be a compact smooth manifold and the associated *-algebra consists of smooth functions on X.

The Gelfand-Naimark Theorem

The algebra A = C(X) of complex-valued functions on a compact topological space X, equipped with the maximum norm

$$|f| = \max_{x \in X} |f(x)|$$

is a commutative C*-algebra [2] . The classical theorem of Gelfand and Naimark characterizes the algebras of the form A = C(X), as commutative unital C*-algebras. This means that for every commutative unital C*-algebra A there exists (up to homeomerphisms) a unique compact topological space X such that $A \cong C(X)$.

As we have earlier observed, the points of the space X are recovered as characters of the associated algebra A. In terms of this identification, the topology on X coincides with the weak*-topology, induced from the dual space A*, consisting of continuous linear functionals on A. It turns out that homomorphisms between C*-algebras are automatically continuous, in particular characters are continuous linear functionals.

THE QUANTUM PLANE

A simple example of quantum geometry is the quantum plane (Figure 1). Usually, a plane is described by two coordinate functions x, y. Naturally, the functions xy and yx are the same since it does not matter whether you measure x first and then y or y first and then x. This is precisely what is lost in the quantum world.

In the quantum plane we replace the property xy = yx by xy = qyx where q is some parameter. We no longer have points, however we can continue to work algebraically with x and y.

Define [x, y] = xy-yx, the commutator bracket. Hence xy = qyx can be rewritten as [x, y] = (q-1)yx

The commutative case is obtained when q = 1.

q-Deformed Calculus

It is interesting to know that one can really do geometry in this setting, where coordinates do not commute. This is the remarkable discovery in recent times. For example, we can follow the approach of Newton-Leibnitz defining differentiation by

$$f'(y) = x^{-1}\left(f(x+y) - f(y)\right)\Big|_{x \to 0}$$

But when $xy \neq yx$ and in particular $xy = qyx$ we get instead

$$\left[f'(x)\right]_q = \frac{f(x) - f(qx)}{(1-q)x}$$

Thus for example the derivative of the function $f(x) = x^n$ in this non-commutative setting would be

$$\left[f'(x)\right]_q = \frac{x^n - q^n x^n}{(1-q)x} = \left(\frac{1-q^n}{1-q}\right)x^{n-1} = \left(1 + q + q^2 + \cdots + q^{n-1}\right)x^{n-1}$$

We observe here that when q=1 the derivative of x^n for the commutative case is obtained, i.e. $f'(x) = nx^{n-1}$.

Basic Notions of q-Calculus

The mathematical study of noncommutative geometry is intimately related to the so-called q-calculus (qnumbers, q-factorials, q-differentials and integrals, basic q-hypergeometric functions, and q-orthogonal polynomials). Here we give a brief introduction to q-numbers and q-factorials which will be required in the subsequent sections.

q-Numbers and q-Factorials

For any nonzero complex number q, the q-number $[a]_q$, $a \in \mathbb{C}$, is defined by

$$[a]_q \equiv [a] := \frac{q^a - q^{-a}}{q - q^{-1}} = \frac{e^{ah} - e^{-ah}}{e^h - e^{-h}} = \frac{\sinh ah}{\sinh h}, \qquad q = \exp h \tag{2.1}$$

We observe that, $\lim_{q \to 1}[a]_q = a$. The following expression which is easily proved will prove useful.

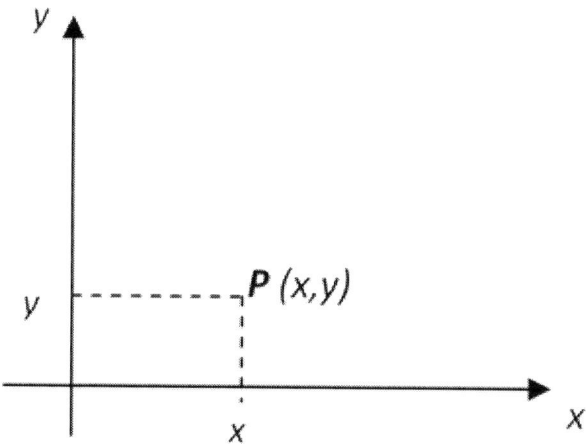

Figure 1: The quantum plane (xy ≠ yx).

$$\llbracket a \rrbracket_q \equiv \llbracket a \rrbracket := \frac{1-q^a}{1-q} = 1+q+q^2+\cdots+q^{a-1} = q^{(a-1)/2}\left[a\right]_{q^{1/2}}$$

(2.2)

Thus given f(x) = x^n, as shown previously $\left[f'(x)\right]_q = \llbracket n \rrbracket x^{n-1}$.

Proposition

The q-numbers satisfy the following relations derived from the property of the exponential function

1) $\left[r\right] = q^{r-1} + q^{r-3} + \cdots + q^{-(r-1)}$

2) $\left[r+s\right] = q^s\left[r\right] + q^{-r}\left[s\right] = q^{-s}\left[r\right] + q^r\left[s\right]$

3) $\left[r-s\right] = q^s\left[r\right] - q^r\left[s\right] = q^{-s}\left[r\right] - q^{-r}\left[s\right]$

4) $\left[r\right]\left[s-t\right] + \left[s\right]\left[t-r\right] + \left[t\right]\left[r-s\right] \equiv 0$

5) $\left[r\right] = \left[2\right]\left[r-1\right] - \left[r-2\right]$

The proof of 1) is easy to see from the fact that;

On q-Deformed Calculus in Quantum Geometry

$$[r] = \frac{q^r - q^{-r}}{q - q^{-1}} = \frac{q^r\left(1 - q^{-2r}\right)}{q\left(1 - q^{-2}\right)} = q^{r-1}\left(\frac{1 - \left(q^{-2}\right)^r}{1 - \left(q^{-2}\right)}\right)$$

Let $u = q^{-2}$, $\therefore$ $\dfrac{1 - \left(q^{-2}\right)^r}{1 - \left(q^{-2}\right)} = \dfrac{1 - u^r}{1 - u} = 1 + u + u^2 + \ldots + u^{r-1}$

Hence;

$$[r] = q^{r-1}\left(1 + q^{-2} + q^{-4} + \cdots + q^{-2(r-1)}\right) = q^{r-1} + q^{r-3} + \cdots + q^{-(r-1)}$$

The rest of the identities can be proved similarly. It is important to note that the relations 1)-5) remain valid when q is considered an indeterminate. Consequently any q-number [n] ,$n \in \mathbb{Z}$, belongs to the space

$\mathbb{Z}\left[q, q^{-1}\right]$ of Laurent polynomials [2] in q with integral coefficients.

Suppose $n \in \mathbb{N}$, then we define the q-factorial $[n]_q! \equiv [n]!$ by setting;

$$[n]! = [1][2]\ldots[n], \qquad [0]! := 1$$

The following expression is quite useful in the theory of hypergeometric functions as well as in combinatorics.

$$\{a : q\}_n := \prod_{j=1}^{n}\left(1 - aq^{j-1}\right) = (1-a)(1-aq)\left(1-aq^2\right)\cdots\left(1-aq^{n-1}\right), \qquad \{a : q\}_0 = 1 \quad (2.3)$$

It is now possible for us to relate the above with the q-factorials. We observe that;

$$[n]! = \frac{q^{-n(n-1)/2}}{\left(1 - q^2\right)^n}\left\{q^2 : q^2\right\}_n$$

$$(2.4)$$

From equation (2.3) we note that $\{q^{-m}:q\}_m = 0, \forall n = m+1, m+2, \dots$ for $|q|<1$, define

$$\{a:q\}_\infty := \lim_{n\to\infty}\{a:q\}_n = \prod_{i=1}^{\infty}\left(1-aq^{j-1}\right) \tag{2.5}$$

Which converges $\forall a \in \mathbb{C}$, and defines an analytic function on $\mathbb{C}$.

Proposition

$$\forall a \in \mathbb{C}, \ |q|<1, \ \{a:q\}_n = \frac{\{a:q\}_\infty}{\{aq^n:q\}_\infty}$$

Proof

$$\frac{\{a:q\}_\infty}{\{aq^n:q\}_\infty} = \frac{\prod_{j=1}^{\infty}\left(1-aq^{j-1}\right)}{\prod_{j=1}^{\infty}\left(1-aq^{n+j-1}\right)}$$

$$= \frac{(1-a)(1-aq)(1-aq^2)(1-aq^3)\cdots(1-aq^{n-1})(1-aq^n)(1-aq^{n+1})\cdots}{(1-aq^n)(1-aq^{n+1})(1-aq^{n+2})\cdots(1-aq^{2n-1})\cdots}$$

$$= (1-a)(1-aq)(1-aq^2)(1-aq^3)\cdots(1-aq^{n-1}) = \{a:q\}_n$$

Remark: We can also define and show that;

$$\{a:q\}_{-n} = \frac{\{a:q\}_\infty}{\{aq^{-n}:q\}_\infty} = \frac{1}{\{aq^{-n}:q\}_n} \tag{2.6}$$

q-Binomial Coefficients

The q-binomial coefficients are defined by the formula;

$$\begin{Bmatrix} n \\ m \end{Bmatrix}_q \equiv \begin{Bmatrix} n \\ m \end{Bmatrix} := \frac{\{q:q\}_n}{\{q:q\}_m \{q:q\}_{n-m}} = \frac{[n]_{q^{1/2}}! \, q^{(n-m)m/2}}{[m]_{q^{1/2}}! [n-m]_{q^{1/2}}!},$$

$$\forall n, m \in \mathbb{N} \cup \{0\}, \; n \geq m. \tag{2.7}$$

Remark: There exist a close analogy between the classical binomial coefficients $\binom{n}{m} = \dfrac{n!}{(n-m)!m!}$ and their q-analogues. Many of the identities satisfied by the former have their counterparts for the q-binomial coefficients. For example the classical identity;

$$\binom{n+1}{m} = \binom{n}{m-1} + \binom{n}{m}$$

Simply translates to;

$$\begin{Bmatrix} n+1 \\ m \end{Bmatrix}_q = \begin{Bmatrix} n \\ m-1 \end{Bmatrix}_q + q^m \begin{Bmatrix} n \\ m \end{Bmatrix}_q = q^{n-m+1} \begin{Bmatrix} n \\ m-1 \end{Bmatrix}_q + \begin{Bmatrix} n \\ m \end{Bmatrix}_q \tag{2.8}$$

Proposition

Let x and y be noncommuting variables satisfying the relation $xy = qyx$, then we have

$$(x+y)^n = \sum_{m=0}^{n} \begin{Bmatrix} n \\ m \end{Bmatrix}_q y^m x^{n-m} = \sum_{m=0}^{n} \begin{Bmatrix} n \\ m \end{Bmatrix}_{q^{-1}} x^m y^{n-m} \tag{2.9}$$

In case q is a primitive p^{th} root of unity, and p is odd, then

$$(x+y)^p = x^p + y^p \tag{2.10}$$

Proof

Equation (2.9) can be established by induction on n, and employing the first identity in (2.8). The second assertion follows directly from (2.9). Observe that:

$$(x+y)^p = \sum_{m=0}^{p} \begin{Bmatrix} p \\ m \end{Bmatrix}_q y^m x^{p-m} = \begin{Bmatrix} p \\ 0 \end{Bmatrix}_q x^p + \begin{Bmatrix} p \\ p \end{Bmatrix}_q y^p + \sum_{m=1}^{p-1} \begin{Bmatrix} p \\ m \end{Bmatrix}_q y^m x^{p-m}$$

From (2.7)

$$\begin{Bmatrix} p \\ m \end{Bmatrix}_q = \frac{\{q:q\}_p}{\{q:q\}_m \{q:q\}_{p-m}}$$

$$= \frac{[p]_{q^{1/2}}! \, q^{(p-m)m/2}}{[m]_{q^{1/2}}! \, [p-m]_{q^{1/2}}!} = 0, \quad m = 1, 2, \ldots, p-1.$$

And $\begin{Bmatrix} p \\ 0 \end{Bmatrix}_q = \begin{Bmatrix} p \\ p \end{Bmatrix}_q = 1.$

Q-Differential and q-Integral Operators

The following are important basic notions derived from their analogue in classical calculus, and will be employed subsequently.

The q-Differential Operator

For $1 \neq q \in \mathbb{C}$, we define the q-differential operator D_q by:

$$D_q f(x) = \frac{f(x) - f(qx)}{(1-q)x}$$

(3.1)

Note that $D_q \to \frac{d}{dx}$ as $q \to 1$.

Proposition:

$$D_q f(x) = \sum_{r=0}^{\infty} \frac{(q-1)^r}{(r+1)!} x^r \frac{d^{r+1}}{dx^{r+1}} f(x), \quad \forall 1 \neq q \in \mathbb{C}$$

(3.2)

Provided the expression on the right hand side exists.

On q-Deformed Calculus in Quantum Geometry

Proof

Let $x+h=qx$, then by Taylor's series, $f(x+h) = f(x) + hf'(x) + \dfrac{h}{2!}f''(x) + \ldots$

$$\therefore \quad f(x+h) - f(x) = hf'(x) + \frac{h}{2!}f''(x) + \cdots$$

$$= \sum_{r=0}^{\infty} \frac{h^{r+1}}{(r+1)!} \frac{d^{r+1}}{dx^{r+1}} f(x)$$

Setting $h = (q-1)x$, we have:

$$D_q f(x) = \frac{f(qx) - f(x)}{(q-1)x}$$

$$= \frac{1}{(q-1)x} \sum_{r=0}^{\infty} \frac{\left[(q-1)x\right]^{r+1}}{(r+1)!} \frac{d^{r+1}}{dx^{r+1}} f(x)$$

$$= \sum_{r=0}^{\infty} \frac{(q-1)^r x^r}{(r+1)!} \frac{d^{r+1}}{dx^{r+1}} f(x)$$

The formula for the product of two functions is given by

$$D_q\{u(x)v(x)\} = \frac{u(x)v(x) - u(qx)v(qx)}{x(1-q)}$$

$$= \frac{v(x)\{u(x) - u(qx)\} + u(qx)\{v(x) - v(qx)\}}{x(1-q)}$$

We can now define the q-analogue of the Newton-Leibnitz rule:

$$D_q\{u(x)v(x)\} = v(x)d_q u(x) + u(qx)d_q v(x) \tag{3.3}$$

Thus,

$$D_q^2 f(x) = D_q\left(D_q f(x)\right) = D_q\left(\sum_{r=0}^{\infty} \frac{(q-1)^r x^r}{(r+1)!} \frac{d^{r+1}}{dx^{r+1}} f(x)\right)$$

$$= \sum_{r=0}^{\infty} \frac{(q-1)^r}{(r+1)!} D_q\left\{x^r \frac{d^{r+1}}{dx^{r+1}} f(x)\right\}$$

$$= \sum_{r=0}^{\infty} \frac{(q-1)^r}{(r+1)!} \left\{ \left(\frac{d^{r+1}}{dx^{r+1}} f(x) \right) D_q x^r + (qx)^r D_q \left(\frac{d^{r+1}}{dx^{r+1}} f(x) \right) \right\}$$

By induction on n, it follows from (3.3) that:

$$D_q^n f(x) = \frac{x^{-n} q^{-n(n-1)/2}}{(q-1)^n} \sum_{j=0}^{n} \begin{bmatrix} n \\ j \end{bmatrix}_q (-1)^j q^{j(j-1)/2} f(q^{n-j} x)$$

$$\tag{3.4}$$

As a special case when n = 2, $D_q^2 f(x)$ is evaluated to give:

$$D_q^2 f(x) = \frac{1}{q(q-1)^2 x^2} \left\{ f(q^2 x) - (1+q) f(qx) + qf(x) \right\}$$

$$\tag{3.5}$$

The q-Integral Operator

The q-integral operator will be defined as the inverse of the q-differential operator.

Given $d_q F(x) = f(x)$, we have:

$$F(x) - F(qx) = (1-q)xf(x)$$

$$\tag{3.6}$$

It then follows that; $F(q^j x) - F(q^{j+1} x) = (1-q)q^j x f(q^j x)$, $j = 0, ..., (n-1)$.

Hence, summing these relations over j gives:

$$F(x) - F(q^n x) = (1-q)x \sum_{r=0}^{n-1} q^r f(q^r x)$$

$$\tag{3.7}$$

Assuming $q \in (0, 1)$ so that $F(q^n x) \to F(0)$ as $n \to \infty$, it follows that

$$F(x) - F(0) = (1-q)x \sum_{r=0}^{\infty} q^r f(q^r x)$$

We can now formally define the q-integral of a function f(x) on a given interval [0, a], as

$$\int_0^\alpha f(x)\,dx_q = \alpha(1-q)\sum_{r=0}^\infty q^r f(q^r\alpha) = \sum_{r=0}^\infty (x_r - x_{r+1})f(x_r),$$

(3.8)

On the semi-infinite interval $[\alpha, \infty]$, the q-integral of a function f(x) is defined as:

$$\int_\alpha^\infty f(x)\,d_q x = \alpha(1-q)\sum_{r=1}^\infty q^{-r} f(q^{-r}\alpha)$$

(3.9)

Over any closed interval [α, b], the q-integral of a function f(x) is given by

$$\int_\alpha^\beta f(x)\,d_q x = \int_0^\beta f(x)\,d_q x - \int_0^\alpha f(x)\,d_q x$$

(3.10)

We now define the integral over the interval $[0, \infty]$. This is achieved by setting $\alpha = 1$ in equations (3.8) and (3.9) and summing to get:

$$\int_0^\infty f(x)\,d_q x = \int_0^1 f(x)\,d_q x + \int_1^\infty f(x)\,d_q x = (1-q)\sum_{r=-\infty}^\infty q^r f(q^r\alpha)$$

(3.11)

The integration by parts formula of Newton-Leibnitz calculus is interpreted in the present noncommutative context as:

$$\int f_2(x)D_q f_1(x)\,d_q x = f_1(x)f_2(x) - \int f_1(qx)D_q f_2(x)\,d_q x$$

APPLICATION

There are a number of applications of the foregoing, we mention here just two, namely:

1.   q-binomial formulae in two variables satisfying a quadratic relation, this has recently been published in [5] and [6]. These rela-

tions have applications in quantum group theory and non-commutative geometry.

2. A recent trend in modern physics is the study of the quantum anti-de Sitter space [7] [8] possibly in connection with q = root of unity [9].

ACKNOWLEDGEMENTS

This work began at the African Institute for mathematical sciences (AIMS) in Muizenberg South Africa, when the first author visited in 2010. I wish to thank the director Prof Fritz Hahne for giving me a postdoctoral fellowship, and for providing a conducive environment which made this research possible.

REFERENCES

1. Connes, A. (1986) Non-Commutative Differential Geometry. Extrait des Publications Mathematiques-IHES, 62. (cited in: Qauntum Principal Bundles and Their Characteristic Classes (pdf), by MICO DURDEVIC, arXiv:q-alg/960505008vi (5 May 1996))
2. Connes A. (1994) Noncommutative Geometry. Academic Press New York.
3. Brateli O. and Robinson, D. (1979) Operator Algebras and Quantum Statistical Mechanics, Volumes 1/2. SpringerVerlag, Berlin.
4. Brown L.G., Douglas, R.G. and Filmore, P.G. (1977) Extensions of C*-Algebras and K-Homology. Annals of Mathematics, 105, 265-324. http://dx.doi.org/10.2307/1970999
5. Benaoum H.B. (1999) (q; h)-Analogue of Newton's Binomial Formula. Journal of Physics A: Mathematical and General, 32, 2037-2040. http://dx.doi.org/10.1088/0305-4470/32/10/019
6. Rosengren H. (1999) Multivariable Orthogonal Polynomials as Coupling Coefficients for Lie and Quantum Algebra Representations. Dissertation, Centre for Mathematical Sciences, Mathematics (Faculty of Science), Lund, 167.
7. Kowalski-Glikman J. (1998) Black Hole Solution of Quantum Gravity. Physics Letters A, 250, 62-66. http://dx.doi.org/10.1016/S0375-9601(98)00706-3
8. Chang Z. (1999) Quantum Anti-De Sitter Space. (reprint)
9. Steinacker H. (1998) Finite Dimensional Unitary Representations of Quantum Antide Sitter Groups at Roots of Unity. Communications in Mathematical Physics, 192, 687-706.http://dx.doi.org/10.1007/s002200050315

CITATION

Maliki, O.S. and Ugwu, E.I. (2014) on q-Deformed Calculus in Quantum Geometry. Applied Mathematics, 5, 1586-1593. http://dx.doi.org/10.4236/am.2014.510151.

Square Root Method

Fernanda Jaiara Dellajustina and Luciano Camargo Martins

Department of Physics, Universidade do Estado de Santa Catarina (UDESC), Joinville, Brazil

4

ABSTRACT

We propose and demonstrate an original geometric argument for the ancient Babylonian square root method, which is analyzed and compared to the Newton-Raphson method. Based on simple geometry and algebraic analysis the former original iterated map is derived and reinterpreted. Time series, fixed points, stability analysis and convergence schemes are studied and compared for both methods, in the approach of discrete dynamical systems.

INTRODUCTION

The oldest known algorithm for successive numerical approximations to the square root of a real number was created by the Babylonians (1950 BC-648 BC) as reported by the Greek mathematician Heron of Alexandria [1] in the first century of our era, named as the Babylonian Method (BABM) in this work. The equivalent mathematical problem has been addressed in the seventeenth century by a more general method to solve numerically the equation $x^2 - k = 0$, the famous Newton-Raphson method (NRM) [2]. In spite of having been studied and applied for centuries, the lack of a geometric argument to the Baby-

Ionian square root method has been a missing link for a better understanding of the ancient Babylonian's mathematics.

Recent applications of iterated maps in numerical analysis have been found in literature, using and extending the techniques of dynamical systems to the study of numerical algorithms and number theory [3] - [5] . Application in technology and hardware devices are also frequent nowadays [6] - [9].

In this work, we study and compare two methods that are based on iterated maps, and some common tools from nonlinear dynamics [10] [11] are used to study the orbits, i.e., the numerical time series obtained for each map are investigated. The numerical approximations $x_i, i = 0, 1, 2, \cdots$ to solve the general equation $x^n = k$, where $n \in N$ and $k \in +R$ are obtained, and the results are valid for positive real values of $n \in +R$. The exact solution is found at the fixed point of each map and its stability is tested. We propose an original geometric argument to construct graphically the BABM basic equation and fill the gap left by this early Babylonian method. We show that both NRM and BABM reduce to a common map for $n = 2$, in spite of having different geometric arguments, and also that the BABM cobwebs are simpler than the NRM ones.

THE NUMERICAL METHODS

From the point of view of discrete dynamic systems BABM is a one-dimensional iterated map defined over the set of real numbers R and over a one-dimensional parameter space (k). The basic idea implemented in this map is that if x is an overestimation for the square root of a real number K, then, k/x will be underestimated and thus the arithmetic mean of these two numbers can be used as best numerical approximation to the exact root. Repeating this procedure with the new value obtained, the approximation can be refined, and so on. The demonstration that this method usually depends on the inequality between arithmetic and geometric mean shows that this average is always an overestimation of the square root, ensuring convergence to the exact root. This algorithm has a quadratic convergence, which means that the number of digits of the numerical approximations nearly doubles at each iteration [12].

The Babylonian method (BABM) is based on the iterated map defined by

$$x_{i+1} = \frac{1}{2}\left[x_i + \frac{k}{x_i}\right]$$

(1.1)

Where k is the radicand and x_i are the successive numerical approximations for the square root of k. The initial condition x_0 is arbitrarily chosen. For example, Table 1 shows the BABM approximations for the square root of 2, i.e., in Equation (1.1) parameter $k = 2$ and $x_0 = 3$. In this case, after six iterations the approximation converges to the exact numerical value of $\sqrt{2}$ within the standard double precision, i.e., to 16 significant digits.

The BABM can be seen as particular case of the more general NRM used to evaluate a zero of the function

$$f(x) = x^2 - k.$$

(1.2)

The geometric construction used by NRM can be reduced to the following geometric path: 1) take an initial value x_i as an approximation of the root of $f(x)$; 2) find the value of the function $f(x_i)$; 3) draw a tangent line from the function at that point using its derivative $f'(x_i)$ at this point; 4) determines the intersection of the tangent with the X-axis, to find the next approximation x_{i+1} to the root; 5) increment i to $i+1$ and return to step 2). This algorithm is graphically illustrated in Figure 1(a).

The numerical approximations generated by the NRM method are in general represented by the iterated map

$$x_{i+1} = x_i - \frac{f(x_i)}{f'(x_i)},$$

(1.3)

Where i indicate the i-th iteration of the map and $f'(x_i)$ am the derivative of the function $f(x)$ at x_i. For the case we are studying the function is used (1.2) and the derivative of this function is $f'(x) = 2x$, assigning these values in Equation (1.3) we obtain

$$x_{i+1} = x_i - \left[\frac{x_i^2 - k}{2x_i} \right] = \frac{1}{2} \left[x_i = \frac{k}{x_i} \right]$$

(1.4)

Which is exactly the same BABM iterated map function.

THE HIDDEN GEOMETRY

There is no historical report showing any geometric argument perhaps used to construct the BABM, but in this section we propose a simple and original one, in the same spirit of that used to construct the NRM. Figure 1(b) shows the schematic geometric path used by the BABM, so that we gain a more intuitive understanding of the convergence schema of this map that permits to demonstrate Equation (1.1).

Table 1: Numerical approximations to $\sqrt{2}$ using $x_0 = 3$ obtained with the Babylonian method

i	xi	i	xi
0	3.000000000000000	4	1.414213780047197
1	1.833333333333333	5	1.414213562373111
2	1.462121212121212	6	1.414213562373095
3	1.414998429894802	7	1.414213562373095

The root of the function (1.2) is found by approximation, from the initial condition, doing arithmetic mean between two numbers. We assume that the root is between two points, do the arithmetic mean between these two points we are closer and closer to the root. These two points are x_i and s_i and the average between them is the next point x_{i+1} in the series, closer to the exact root $\sqrt{k}$. So, from the initial condition x_i, the auxiliary point s_i is found, and their average renders the next point

The Hidden Geometry of the Babylonian Square Root Method

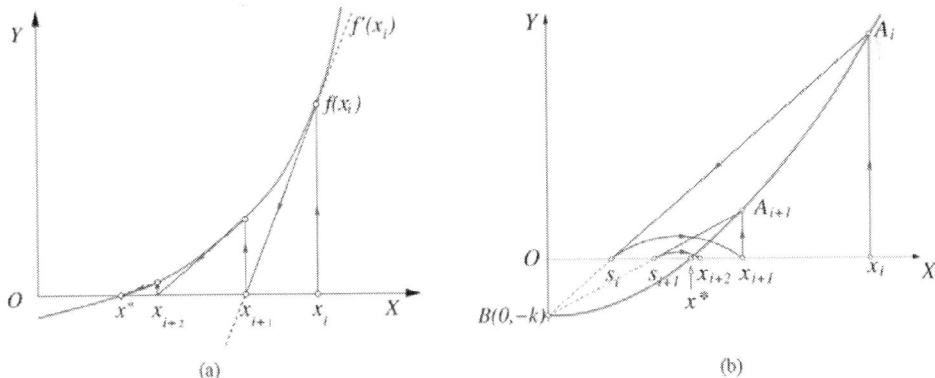

Figure 1: (a) The geometric paths for the Newton-Raphson and (b) Babylonian methods.

x_{i+1} of the map series, as shown in Figure 1(b). This point is obtained by knowing the equation of the auxiliary straight line $\overline{A_iB}$ that intersects the X-axis at the auxiliary point s_i. This line contains the auxiliary point $B(0,-k)$, vertex of the parabola.

Given an initial condition we start the geometric path construction. The first step is to find out the equation of the first auxiliary line $\overline{A_iB}$, whose slope is $m_i = \Delta y_i / \Delta x_i$. According to Figure 1(b), $\Delta y_i = f(x_i)+k$ and $\Delta x_i = x_i$, and the slope is $m_i = (f(x_i)+k)/x_i$, and replacing $f(x_i)$ for their function (1.2) we find the slope,

$$m_i = \frac{(x_i^2 - k)+k}{x_i} = x_i, i = 0,1,2,3,L,N$$

(1.5)

That generates the x_i series recursively.

Drawing the straight line $y = mx+b$ whose linear coefficient is $-k$ passing through the point $B(0,-k)$, we have $y = x_ix - k$, and the auxiliary point s_i can be found when $y = 0$, which is the intersection point of the line with the X-axis, and solving $0 = x_is_i - k$ we finally find the auxiliary points $s_i = k / x_i$, for $i = 0,1,2,3,\cdots,N$, and the original Equation (1.1) of BABM is exactly recovered. The term k/x_i corresponds to the auxiliary points$_i$, and the BABM basically works by doing the arithmetic mean between these points.

The convergence analysis of time series generated by BABM will be done analytically and graphically by determining the it's fixed point and testing its stability. By inspecting Equation (1.1) we see its general form $x_{i+1} = f(x_i)$, with the mapping function

$$f(x) = \frac{1}{2}\left(x + \frac{k}{x}\right)$$

(1.6)

Where k is a fixed parameter that will be analyzed below.

In general, the first values of the series $\{x_i i = 0, 1, 2, \cdots\}$ are irregular, not having a well-defined pattern, and this irregular initial behavior is called the transient. After discarding the transient, the values of x_i can basically: 1) cycle between N fixed values, or a period-N orbit; 2) assume an aperiodic bounded sequence of values that are never repeated, or a chaotic orbit; 3) an unbound orbit, or divergence.

When a series converges asymptotically to a single fixed value x^*, the fixed point, we have an orbit of period-1. An attractive fixed point of a function $f(x)$ is a fixed point x^* of $f(x)$ such that for any value of x in the domain that is close enough to x^*, the iterated function sequence $x, f(x), f(f(x)), f(f(f(x))), L$ converges to x^*. An expression of prerequisites and proof of the existence of such solution is given by Banach's fixed point theorem [13]. An attractive fixed point is also called a stable fixed point. However, if the map $f(x)$ is continuously differentiable in an open neighbourhood of a fixed point x^*, the stability criterion (1.10) is satisfied.

FIXED POINTS AND STABILITY ANALYSIS

For the analytical determination of a fixed point x^*, we have to solve the equation $x_{i+1} = x_i$, or

$$x^* = f(x^*),$$ (1.7)

That for BABM is

$$x^* = \frac{1}{2}\left(x^* + \frac{k}{x^*}\right)$$ (1.8)

Whose solution for x^* renders

$$x^* = \pm\sqrt{k}$$ (1.9)

And the existence of this fixed points is the starting point to use the map for square root extraction. For the stability of the fixed point of the map, we have to ensure that

$$\left|f'(x^*)\right| < 1$$ (1.10)

Which is the general condition for stability of fixed points of any onedimensional map [14]. Since $f'(x^*)$ is the derivative of the function (1.6), we have,

$$f'(x) = \frac{1}{2}\left(1 - \frac{k}{x^2}\right)$$ (1.11)

And according to the stability criterion (1.10) that at the point fixed (1.9) is

$$f'(x^*) = \frac{1}{2}\left(1 - \frac{k}{(\pm\sqrt{\ })^2}\right) = 0,$$ (1.12)

And its fixed points $\pm\sqrt{k}$ are both stable, regardless of the value of k, so the stability criterion is verified with no dependence on this parameter.

Another important tool to analyze the orbit of a map and its evolution in time is called return diagram or cobweb. A cobweb is built with all the values of x_i obtained in series to construct a graph that has coordinates x_i to the X-axis and x_{i+1} for the Y-axis, recursively following the steps: 1) choose an initial condition x_0 and iterate the map to obtain the next point $x_1 = f(x_0)$; 2) draw the line segment from $(x_0, 0)$ to (x_0, x_1); 3) join this point to point (x_1, x_1) on the identity line y=x; 4) use this point to return to X-axis, joining it to point $(x_1, 0)$; 5) go to to step 2).

Figure 2(a) shows the BABM cobweb for the square root of 2. Plotted on the graph is the equation of BABM, the identity function and the return diagram, for the initial condition $x_0 = 9$. Figure 2(b) plots separately the linear and nonlinear terms of Equation (1.4) indicating that when the linear term has a weight greater than he nonlinear one the map converges to the root, independent of the values of parameters n and. For parameter $k = 2$, Figure 2(b) shows in red the linear term x/2 in NRM, in green the nonlinear term k/n and in blue the sum of both terms, according Equation (1.4). The identity function is drawn in black and yellow is used for The tangent line at the fixed point of the map. The linear term has a greater weight ensuring that $|f'(x^*)| < 1$, what guarantees that the fixed point is stable. Other important information we can obtain from this figure is the stability of the fixed point, since the map function has a minimum exactly at fixed point, and thus the derivative of the map is zero at this point. According to the stability criterion this is necessary and sufficient condition for the fixed point to be stable.

CONCLUSIONS

The use of iterated maps to solve the fundamental mathematical problem of square root estimation by numerical approximations was revisited and some tools from nonlinear dynamics were used to predict their stable fixed points and test the behavior of the corresponding time series over a large region of parameter k.

The Hidden Geometry of the Babylonian Square Root Method

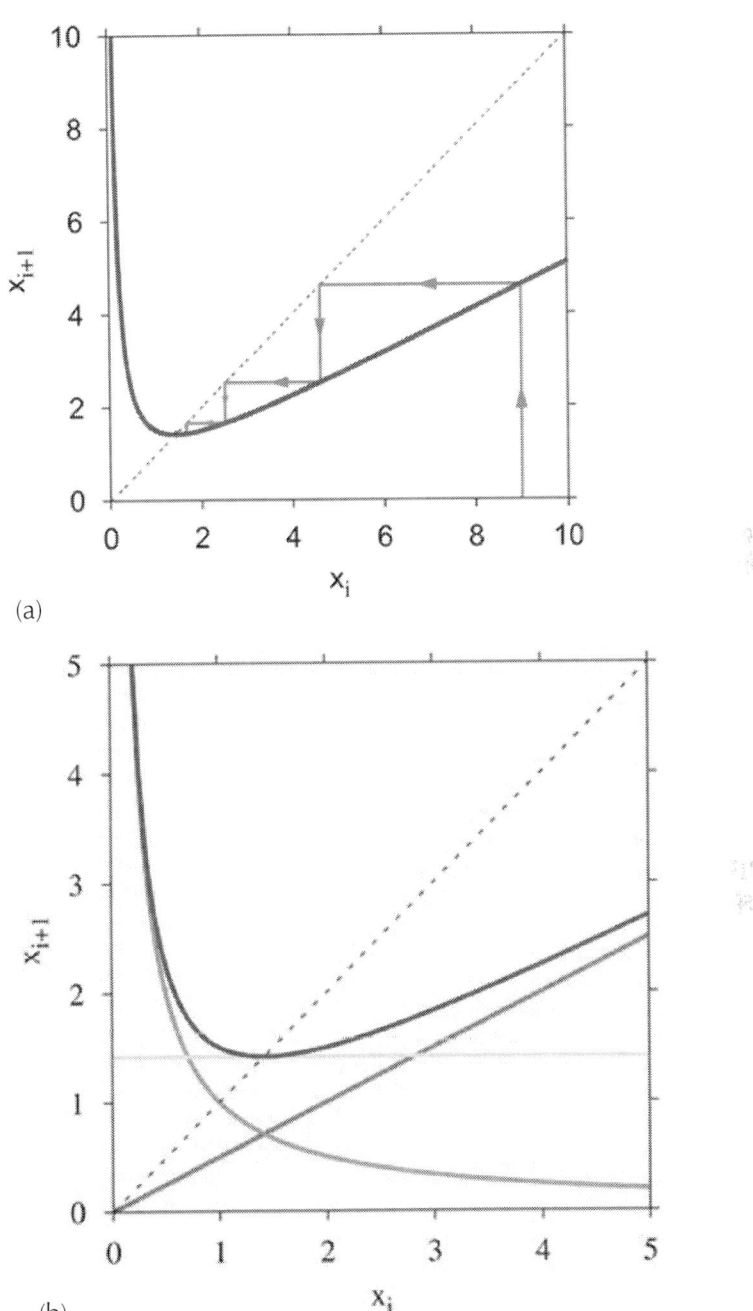

(a)

(b)

Figure 2: For $k = 2$ (a) the BABM cobweb for $X_0 = 9$; (b) the NRM linear term (red), nonlinear term (green) and the map function (blue).

The main result of this paper is fulfilled once we have proposed and demonstrated an original geometric argument to the underlying geometry in the Babylonian square root method, the oldest known and one of the most efficient methods to solve this classical and current problem. The proposed argument is very simple and intuitive, and can be easily extended to other similar maps, and perhaps its basic idea could be useful for constructing new iterated maps, from the geometrical point of view.

ACKNOWLEDGEMENTS

This work was partially supported by the Brazilian agency Conselho Nacional de Desenvolvimento Cientfico e Tecnológico—CNPq and Universidade do Estado de Santa Catarina—UDESC.

REFERENCES

1. Heath, T. (1923) A History of Greek Mathematics. The Mathematical Gazette, 11, 348-351. http://dx.doi.org/10.2307/3602335
2. Verbeke, J. and Cools, R. (1995) the Newton-Raphson Method. International Journal of Mathematical Education in Science and Technology, 26, 177-193. http://dx.doi.org/10.1080/0020739950260202
3. Faber, X. and Voloch, J.F. (2011) On the Number of Places of Convergence for Newton's Method over Number Fields. Journal de Théorie des Nombres de Bordeaux, 23, 387-401.
4. Grau-Sánchez, M. and Daz-Barrero, J.L. (2011) A Technique to Composite a Modified Newton's Method for Solving Nonlinear Equations. ArXiv e-prints.
5. Pan, B., Cheng, P. and Xu, B. (2005) In-Plane Displacements Measurement by Gradient-Based Digital Image Correlation. SPIE Proceedings, 5852, 544-551.
6. Amin, A.M., Thakur, R., Madren, S., Chuang, H.-S., Thottethodi, M., Vijaykumar, T., Wereley, S.T. and Jacobson, S.C. (2013) Software-Programmable Continuous-Flow Multi-Purpose Lab-on-a-Chip. Microfluidics and Nanofluidics, 15, 647-659. http://dx.doi.org/10.1007/s10404-013-1180-2
7. Mungan, C.E. and Lipscombe, T.C. (2012) Babylonian Resistor Networks. European Journal of Physics, 33, 531. http://dx.doi.org/10.1088/0143-0807/33/3/531
8. Senthilpari, C., Mohamad, Z.I. and Kavitha, S. (2011) Proposed Low Power, High Speed Adder-Based 65-nm Square Root Circuit. Microelectronics Journal, 42, 445-451. http://dx.doi.org/10.1016/j.mejo.2010.10.015

9. Sun, T., Tsuda, S., Zauner, K.-P. and Morgan, H. (2010) On-Chip Electrical Impedance Tomography for Imaging Biological Cells. Biosensors and Bioelectronics, 25, 1109-1115. http://dx.doi.org/10.1016/j.bios.2009.09.036

10. Ausloos, M. and Dirickx, M. (2005) The Logistic Map and the Route to Chaos: From the Beginnings to Modern Applications. Springer, New York.

11. Eve, J. (1963) Starting Approximations for the Iterative Calculation of Square Roots. The Computer Journal, 6, 274- 276. http://dx.doi.org/10.1093/comjnl/6.3.274

12. Macleod, A.J. (1984) A Generalization of Newton-Raphson Method. International Journal of Mathematical Education in Science and Technology, 15, 117-120. http://dx.doi.org/10.1080/0020739840150116

13. Banach, S. (1992) Sur les oprations dans les ensembles abstraits et leur application aux quations intgrales. Fundamenta Mathematicae, 3, 133-181.

14. Lyapunov, A.M. (1992) the General Problem of the Stability of Motion. International Journal of Control, 55, 531-534. http://dx.doi.org/10.1080/00207179208934253

CITATION

Dellajustina, F. and Martins, L. (2014) the Hidden Geometry of the Babylonian Square Root Method. Applied Mathematics, 5, 2982-2987. doi: 10.4236/am.2014.519284.

On the Cozero-Divisor Graphs of Commutative Rings

Mojgan Afkhami and
Kazem Khashyarmanesh
Department of Pure Mathematics, Ferdowsi
University of Mashhad, Mashhad, Iran

ABSTRACT

Let R be a commutative ring with non-zero identity. The cozero-divisor graph of R, denoted by $\Gamma'(R)$, is a graph with vertices in $W^*(R)$, which is the set of all non-zero and non-unit elements of R, and two distinct vertices a and b in $W^*(R)$ are adjacent if and only if $a \notin bR$ and $b \notin bR$. In this paper, we investigate some combinatorial properties of the cozero-divisor graphs $\Gamma'(R[x])$ and $\Gamma'(R[[x]])$ such as connectivity, diameter, girth, clique numbers and planarity. We also study the cozero-divisor graphs of the direct products of two arbitrary commutative rings.

INTRODUCTION

Let R be a commutative ring with non-zero identity and let Z(R) be the set of zero-divisors of R. For an arbitrary subset A of R, we put $A^* = A \setminus \{0\}$. The zero-divisor graph of R, denoted by $\Gamma'(R)$, is an undirected graph whose vertices are elements of $Z^*(R)$ with two distinct vertices a and b are adjacent if and only if ab = 0.

The concept of zero-divisor graph of a commutative ring was introduced by Beck [1], but this work was mostly concerned with colorings of rings. The above definition first appeared in Anderson and Livingston

[2], which contained several fundamental results concerning the graph $\Gamma(R)$. The zero-divisor graphs of commutative rings have been studied by several authors. For instance, the preservation and lack thereof of basic properties of $\Gamma(R)$ under extensions to rings of polynomials and power series was studied by Axtell, Coykendall and Stickles in [3] and Lucas in [4]. Also Axtell, Stickles and Warfel in [5], considered the zero-divisor graphs of direct products of commutative rings.

Let $W(R)$ be the set of all non-unit elements of R. For an arbitrary commutative ring R, the cozero-divisor graph of R, denoted by $\Gamma'(R)$, was introduced in [6], which is a dual of zero-divisor graph $\Gamma(R)$ "in some sense". The vertex-set of $\Gamma'(R)$ is $W^*(R)$ and for two distinct vertices a and b in $W^*(R)$, a is adjacent to b if and only if $a \notin bR$ and $b \notin bR$, where cR is an ideal generated by the element c in R. Some basic results on the structure of this graph and the relations between two graphs $\Gamma(R)$ and $\Gamma'(R)$ were studied in [6].

In this paper, we study the cozero-divisor graphs of the rings of polynomials, power series and the direct product of two arbitrary commutative rings. Also, we look at the preservation of the diameter and girth of the cozero-divisor graphs in some extension rings. Our results "in some sense" are the dual of the main results of [3-5].

Throughout the paper, R is a commutative ring with non-zero identity. We denote the set of maximal ideals and the Jacobson radical of R by $\text{Max}(R)$ and $J(R)$, respectively. Also, $U(R)$ is the set of all unit elements of R. By a local ring, we mean a (not necessarily Noetherian) ring with a unique maximal ideal.

In a graph G, the distance between two distinct vertices a and b, denoted by $d_G(a,b)$, is the length of the shortest path connecting a and b, if such a path exists; otherwise, we set $d(a,b) := \infty$. The diameter of a graph G is $\text{diam}(G) = \sup\{d_G(a,b) \mid a \text{ and } b \text{ are distinct vertices of } G\}$

The girth of G, denoted by $g(G)$, is the length of the shortest cycle in G, if G contains a cycle; otherwise, $g(G) := \infty$. Also, for two distinct vertices a and b in G, the notation a-b means that a and b are adjacent. A graph G is said to be connected if there exists a path between any two distinct vertices, and it is complete if it is connected with diameter one. We use K_n to denote the complete graph with n vertices. Moreover, we say that G is totally disconnected if no two vertices of G are adjacent. For a graph G, let $\chi(G)$ denote the chromatic number of the graph G, i.e., the minimal number of colors which can be assigned to the vertices of G in such a way that any two adjacent vertices have different colors. A clique of a graph is any complete subgraph of the graph and the number of vertices in a largest clique of G, denoted by clique(G), is called the clique number of G. Obviously $\chi(G) \geq \text{clique}(G)$ (cf. see [7, p. 289]). For a positive integer r, an r-partite graph is one whose vertex-set can be partitioned into r subsets so that no edge has both ends in any one subset. A complete r-partite graph is one in which each vertex is joined to every vertex that is not in the same subset. The complete bipartite graph (2-partite graph) with subsets containing m and n vertices, respectively, is denoted by $K_{m,n}$. A graph is said to be planar if it can be drawn in the plane so that its edges intersect only at their ends. A subdivision of a graph is any graph that can be obtained from the original graph by replacing edges by paths. A remarkable simple characterization of the planar graphs was given by Kuratowski in 1930. Kuratowski's Theorem says that a graph is planar if and only if it contains no subdivision of K_5 or $K_{3,3}$ (cf. [8, p. 153]). Also, the valency of a vertex a is the number of edges of the graph G incident with a.

COZERO-DIVISOR GRAPH OF $R[x]$

In this section, we are going to study some basic properties of the cozero-divisor graph of the polynomial ring $R[x]$. To this end, we first gather together the well-known properties of the polynomial ring $R[x]$, which are needed in this section.

Remarks 2.1: Let $f(x) = \sum_{i=0}^{n} a_i x^i$ be an arbitrary element in $R[x]$. Then we have the following statements:

- $f(x)$ is a unit in $R[x]$ if and only if a_0 is a unit and the coefficients $a_1, ..., a_n$ are nilpotent elements of R.

- $f(x)$ is nilpotent if and only if the coefficients $a_1, ..., a_n$ are nilpotent.

- $J(R[x]) = \text{Nil}(R[x])$, where $\text{Nil}(R[x])$ is the nilradical of $R[x]$.

- Since the polynomials x and $1 + x$ are non-units, $R[x]$ is a non-local ring.
- By part (i), it is easy to see that $\Gamma'(R)$ is an induced subgraph of $\Gamma'(R[x])$.

In the following theorem, we show that $\Gamma'(R[x])$ is always connected and its diameter is not exceeding three.

Theorem 2.2: The graph $\Gamma'(R[x])$ is connected and $\mathrm{diam}(\Gamma'(R[x])) \leq 3$.

Proof: Since $R[x]$ is a non-local ring, by [1, Theorem 2.5], it is enough to show that, for every non-zero element $f(x) \in J(R[x])$, there exist $m \in \mathrm{Max}(R[x])$ and $g(x) \in m \setminus J(R[x])$ such that $f(x) \notin g(x)(R[x])$. Now, assume that $f(x)$ is a non-zero polynomial in $J(R[x])$ of degree t. Since x is a non-unit element in $R[x]$, there exists a maximal ideal m of $R[x]$ such that $x \in m$. So $x^{t+1} \in m$. On the other hand, by parts (ii) and (iii) of Remarks 2.1, $x^{t+1} \notin J(R[x])$. Also $f(x) \notin x^{t+1}R[x]$. Hence the graph $\Gamma'(R[x])$ is connected and $\mathrm{diam}(\Gamma'(R[x])) \leq 3$.

The following proposition states that the diameter of $\Gamma'(R[x])$ is never one.

Proposition 2.3: The graph $\Gamma'(R[x])$ is never complete.

Proof: Clearly $x \in W^*(R[x])$ and $\{0, x, x^2\} \subseteq xR[x]$. The claim now follows from [1, Theorem 2.1].

The following corollary is an immediate consequence of Theorem 2.2 and Proposition 2.3.

Corollary 2.4: $\mathrm{diam}(\Gamma'(R[x])) = 2$ or 3.

Proposition 2.5: Suppose that $J(R[x]) = 0$. Then $\mathrm{diam}(\Gamma'(R[x])) = 2$. In particular, if R is reduced, then $\mathrm{diam}(\Gamma'(R[x])) = 2$.

Proof: In view of [1, Corollary 2.4], $\mathrm{diam}(\Gamma'(R[x])) \leq 2$. Now, by Proposition 2.3, one can conclude that $\mathrm{diam}(\Gamma'(R[x])) = 2$. Also, if R is reduced, then by Remarks 2.1 (ii), $\mathrm{Nil}(R[x]) = 0$ and so, by Remarks 2.1 (iii), $J(R[x]) = \mathrm{Nil}(R[x]) = 0$.

In the next two theorems, we investigate the girth of the graph $\Gamma'(R[x])$.

Theorem 2.6: Suppose that R is a non-reduced ring. Then every element of $J(R[x])$ is in a cycle of length three.

Proof: Assume that $f(x)$ is a non-zero element in $J(R[x])$ of degree n and consider the elements x^t and $1 + x^t$ in $R[x]$, where $t > n$. Then, by Remarks 2.1 (iv), there exist maximal ideals m_1 and m_2 of $R[x]$ such that $x^t \in m_1 \setminus m_2$ and $1 + x^t \in m_2 \setminus m_1$. Since $t > n$, $f(x) \notin x^t R[x]$ and $f(x) \notin (1 + x^t)R[x]$. Also, by parts (ii) and (iii) of Remarks 2.1, x^t and $1 + x^t$ do not belong to $J(R[x])$. So, $x^t \notin f(x)R[x]$ and $(1 + x^t) \notin f(x)R[x]$. Thus, $f(x)$ is adjacent to both distinct vertices x^t and $1 + x^t$. Moreover, it is easy to see that x^t is adjacent to $1 + x^t$. Therefore we have the cycle $f(x) - x^t - (1 + x^t) - f(x)$.

Theorem 2.7: $g(\Gamma'(R[x])) = 3$.

Proof: Consider the elements $x, 1 + x$ and $1 + x + x^2$ in $R[x]$. So there exist two maximal ideals m_1 and m_2 in $R[x]$ such that $x \in m_1 \setminus m_2$ and $1 + x \in m_2 \setminus m_1$. Hence the vertex x is adjacent to $1 + x$. Also, clearly $x \notin (1 + x + x^2)R[x]$ and $1 + x \notin (1 + x + x^2)R[x]$. Now, since the polynomials x and $1 + x$ do not divide the polynomial $1 + x + x^2$, we have that $1 + x + x^2 \notin xR[x]$ and $1 + x + x^2 \notin (1 + x)R[x]$. Hence $(1 + x + x^2) - (1 + x) - x - (1 + x + x^2)$ is the required cycle. In the next theorem we study the clique number of $\Gamma'(R[x])$.

Theorem 2.8: In the graph $\Gamma'(R[x])$, clique$(\Gamma'(R[x]))$ is infinity and hence the chromatic number $\chi(\Gamma'(R[x]))$ is infinity.

Proof: Let n be a positive integer and consider the subgraph G_n of $\Gamma'(R[[x]])$ with vertex-set $\{1 + x + \cdots + x^i \mid = n, \cdots, 2n\}$. Now, for every two distinct polynomials $1 + x + \cdots + x^i$ and $1 + x + \cdots + x^j$ with $i < j$, clearly we have that $1 + x + \cdots + x^i \notin (1 + x + \cdots + x^j)R[x]$. Also, since $n \le i, j \le 2n$, we have that $j - 2i < 1$. This means that $j - i - 1 < i$, and so $1 + x + \cdots + x^i$ does not divide the polynomial $1 + x + \cdots + x^j$. Thus $1 + x + \cdots + x^j \notin (1 + x + \cdots + x^i)R[x]$. Hence, G_n is a complete subgraph of $\Gamma'(R[x])$ which is isomorphic to k_{n+1}. So clique$(\Gamma'(R[x]))$ is infinity. This implies that $\chi(\Gamma'(R[x]))$ is infinity.

Theorem 2.9: The cozero-divisor graph $\Gamma'(R[x])$ is not planar.

Proof: In view of the proof of Theorem 2.8, for all positive integers n, the cozero-divisor graph $\Gamma'(R[x])$ has a complete subgraph isomorphic to K_n. In particular, for n=5, the graph K_5 is a subgraph of $\Gamma'(R[x])$. So, by Kuratowski's Theorem (cf. [8, p. 153]), $\Gamma'(R[x])$ is not planar.

Recall that a graph on n vertices such that n−1 of the vertices have valency one, all of which are adjacent only to the remaining vertex a, is called a star graph with center a. Also, a refinement of a graph H is a graph G such that the vertex-sets of G and H are the same and every edge in H is an edge in G. Now, we have the following result.

Proposition 2.10: If there exists a maximal ideal m of R with $|m|= 2$, then there is a refinement of a star graph in the structure of $\Gamma'(R[x])$.

Proof: Suppose that $m = \{0,a\}$ is a maximal ideal of R. Then, for every element $b \in W^*(R)$ with $a \neq b$, we have that $b \notin aR$. Also $a \notin bR$. Hence, a is adjacent to b. Therefore, $\Gamma'(R)$ is a refinement of a star graph with center a. Now, by Remarks 2.1 (v), $\Gamma'(R)$ is an induced subgraph of $\Gamma'(R[x])$. So $\Gamma'(R[x])$ contains a refinement of a star graph.

COZERO-DIVISOR GRAPH OF $R[[x]]$

We begin this section with some elementary remarks about the rings of power series which may be valuable in turn. These facts can be immediately gained from the elementary notes about power series.

Remarks 3.1

- $f(x) = \sum_{i=0}^{\infty} a_i x^i$ is a unit in $R[[x]]$ if and only if a_0 is a unit in R.

- $f(x) = \sum_{i=0}^{\infty} a_i x^i$ belongs to the Jacobson radical of $R[[x]]$ if and only if a_0 belongs to the Jacobson radical of R.

- $R[[x]]$ is a local ring if and only if R is local.

- The cozero-divisor graph $\Gamma'(R)$ is an induced subgraph of $\Gamma'(R[[x]])$, but $\Gamma'(R[x])$ is not a subgraph of $\Gamma'(R[[x]])$, since $1 + x$ is a vertex of $\Gamma'(R[x])$ but it is not in the vertex-set of $\Gamma'(R[[x]])$.

In the following proposition, we study the connectivity and diameter of $\Gamma'(R[[x]])$, whenever R is non-local.

Proposition 3.2: Let R be a non-local ring. Then the cozero-divisor graph $\Gamma'(R[[x]])$ is connected and $\mathrm{clique}(\Gamma'(R[x])) \le 3$.

Proof: Suppose that $f(x) = \sum_{i=0}^{\infty} a_i x^i$ is a non-zero element in $J(R[[x]]))$. By [6, Theorem 2.5], it is enough to show that $f(x)$ is adjacent to some element in $W(R[[x]])) \setminus J(R[[x]]))$. In this regard, we have the following two cases:

Case 1: Assume that $a_i \in U(R)$, for some $i \ge 1$ and consider an element b in $W(R) \setminus J(R)$. We will show that $f(x)$ is adjacent to b. Clearly, by Remarks 3.1 (i)(ii), $b \in W(R[[x]])) \setminus J(R[[x]]))$. Now, assume in contrary that $b \in (x)R[[x]]$ and look for a contradiction.

We have that $b = f(x)g(x)$, for some $g(x) = \sum_{i=0}^{\infty} g_i x^i$ in $R[[x]]$. Since, by Remarks 3.1 (ii), $a_0 \in J(R)$, we have that $b \in J(R)$ which is impossible. Also, if $f(x) \in bR[[x]]$, then $a_i = bg_i$, for some non-zero element g_i in R, which is impossible. Therefore $f(x)$ and b are adjacent.

Case 2: Suppose that $a_i \in W(R)$, for all $i \ge 1$. First assume that $a_i \in J(R)$, for some $i \ge 1$. Hence, there exist maximal ideals m and m' such that $a_i \in m \setminus m'$. By considering an element b in $m \setminus m'$, one can conclude that a_i is adjacent to b. Now if $f(x) \in bR[[x]]$, then $a_i = bg_i$, for some non-zero element g_i in R which is a contradiction because the vertices a_i and b are adjacent. On the other hand, $b \notin f(x)R[[x]]$. Thus the vertices $f(x)$ and b are adjacent.

Now, let $a_i \in J(R)$, for all $i \ge 1$. Choose $b \in W(R) \setminus J(R)$. Hence $b \in W(R[[x]])) \setminus J(R[[x]]))$. We claim that $f(x)$ is adjacent to $b + x^{t+1}$, where t is the least non-zero power of x in the polynomial $f(x)$. Clearly, $f(x) \notin (b + x^{t+1})R[[x]]$. Now, if $b + x^{t+1} = f(x)g(x)$ for some $g(x) = \sum_{i=0}^{\infty} g_i x^i$ in $R[[x]]$, then $1 = a_0 g_{t+1} + \ldots\ldots + a_{t+1}g_0$ which belongs to $J(R)$ and this is impossible. Hence, we have that $f(x)$ is adjacent to $b + x^{t+1}$

Therefore $\Gamma'(R[[x]])$ is connected and also, by considering the above cases, it is routine to check that $\text{diam}(\Gamma'(R[[x]])) \leq 3$.

In the next lemma, we investigate the adjacency in $\Gamma'(R[[x]])$ in the case that R is a local ring.

Lemma 3.3: Assume that R is a local ring with maximal ideal m. Let $f(x) = \sum_{i=0}^{\infty} a_i x^i$ be a non-zero element in $J(R[[x]])$. Then we have the following statements:

- if $a_0 \neq 0$, then $f(x)$ is adjacent to x;
- if $a_0 \neq 0$ and $a_i \in U(R)$, for some $i \geq 1$, then $f(x)$ is adjacent to all non-zero elements of $J(R)$; and
- if $a_0 = 0$ and $a_i \in m$, for all $i \geq 1$, then $f(x)$ is adjacent to x^{t+1}, where t is the least non-zero power of x in $f(x)$.

Proof

1) Assume on the contrary that $x = f(x)g(x)$, where $g(x) = \sum_{i=0}^{\infty} g_i x^i$. Then $a_0 g_0 = 0$ and $1 = a_0 g_1 + g_0 a_1$. Since a_0 and g_0 belong to m, we have that $1 \in m$ which is a contradiction. Also we have that $f(x) \notin xR[[x]]$. Thus $f(x)$ is adjacent to x.
2) Let b be a non-zero element in m. Then if $f(x) \in bR[[x]]$, we conclude that $a_i \in m$ which is impossible. So $f(x) \notin bR[[x]]$. Now, since $a_0 = 0$ $b \notin f(x)R[[x]]$. Therefore, $f(x)$ is adjacent to b.
3) Clearly, since t is the least non-zero power of x in $f(x)$, $f(x) \notin x^{t+1}R[[x]]$. Moreover, if $x^{t+1} \notin f(x)g(x)$, for some $g(x) = \sum_{i=0}^{\infty} g_i x^i$ in $R[[x]]$, then $1 = a_0 g_{t+1} + \cdots + a_{t+1} g_0$

This means that $1 \in m$ which is impossible. Hence $x^{t+1} \notin f(x)R[[x]]$. So $f(x)$ is adjacent to x^{t+1}.

The following result, which is one of our main results in this section, states that $\Gamma'(R[[x]])$ is connected and the diameter of $\Gamma'(R[[x]])$ is not exceeding four.

Theorem 3.4: The cozero-divisor graph $\Gamma'(R[[x]])$ is always connected and also $\text{diam}(\Gamma'(R[[x]])) \leq 4$.

Proof: Owing to Proposition 3.2, the result holds in the case that R is non-local. Assume that R is a local ring with maximal ideal m.

In view of part (iii) of Remarks 3.1, $R[[x]]$ is also a local ring. Now, let $f(x) = \sum_{i=0}^{\infty} a_i x^i$ and $g(x) = \sum_{i=0}^{\infty} g_i x^i$ be two non-zero elements in $W(R[[x]])$ s that are not adjacent. We have the following cases for consideration:

Case 1: $a_0 \neq 0$ and $b_0 \neq 0$. Then by Lemma 3.3 (i), we have that $f(x) - x - g(x)$.

Case 2: $a_0 = 0 = b_0$. If $a_i, b_j \in U(R)$, for some i, j, then by part (ii) of Lemma 3.3, $f(x) - c - g(x)$, for all non-zero elements c in m.

Also, if $a_i, b_j \in m$, for all i,j, and t, t' are the least non-zero powers of x in $f(x)$ and, $g(x)$ respectively, with $t' \leq 1$, then by Lemma 3.3 (iii), one can easily check that $f(x) - x^{t+1} - g(x)$.

Finally, we may assume that for some positive integer i, $a_i \in U(R)$ and $b_j \in m$, for all j. Thus, by parts (ii) and (iii) of Lemma 3.3, we have the path $f(x) - x - c - x^{t+1} - g(x)$, where c is a non-zero element in m and t is the least non-zero power of x in $g(x)$.

Case 3: Without loss of generality, we may assume that $a_0 \neq 0$ and $b_0 \neq 0$. So, if $b_j \in U(R)$, for some j, then in view of parts (i) and (ii) of Lemma 3.3, we have the path $f(x) - x - c - g(x)$, where c is a non-zero element in m.

Moreover, if $b_j \in U(R)$, for all j, then by Lemma 3.3, we have $f(x) - x - c - x^{t+1} - g(x)$, where c a non-zero element in m and t is the least non-zero power of x in $g(x)$.

Therefore, the cozero-divisor graph $\Gamma'(R[[x]])$ is connected and in view of the above cases, one can easily check that $diam(\Gamma'(R[[x]])) \leq 4$.

The following lemma is needed in the sequel.

Lemma 3.5: Let $a \in w^*(R)$ and let i and j be positive integers such that $i < j < 2i$. Then the vertices $a + x^i$ and $a + x^j$ are adjacent in $\Gamma'(R[[x]])$.

Proof: Suppose to the contrary that $a + x^j = (a + x^i)f(x)$, where $f(x) = b_0 + b_1 x + \ldots \ldots + b_{j-i} x^{j-i}$ is a non-zero polynomial in $R[[x]]$. So, we have $ab_0 = a$ and $b_0 = 0$. Thus $a = 0$ which is a contradiction. Hence $a + x^j \notin (a + x^i)R[[x]]$. Also, clearly $a + x^i \notin (a + x^j)R[[x]]$. So the vertices $a + x^i$ and $a + x^j$ are adjacent in the cozero-divisor graph $\Gamma'(R[[x]])$. In the next theorem, we show that $g(\Gamma'(R[[x]])) = 3$.

Theorem 3.6: The cozero-divisor graph $\Gamma'(R[[x]])$ has girth three.

Proof: Let $a \in W * (R)$. Consider the elements x, $a + x^2$ and $a + x^3$ in $R[[x]]$. Clearly, $x \notin (a + x^2)(R)[[x]]$ and $x \notin (a + x^3)(R)[[x]]$. Also, since $a \neq 0$, $a + x^2$ and $a + x^3$ don't belong to $xR[[x]]$. Hence, we have the following path $(a + x^2) - x - (a + x^3)$.

Now, in view of Lemma 3.5, one can conclude that $a + x^2$ and $a + x^3$ are adjacent. Therefore, we have the cycle $x - (a + x^2) - (a + x^3) - x$. Hence, $g(\Gamma'(R[[x]])) = 3$.

In the next theorem, we compute the clique number of $\Gamma'(R[[x]])$.

Theorem 3.7: In the graph $\Gamma'(R[[x]])$, clique$(\Gamma'(R[[x]]))$ is infinity and hence $\chi(\Gamma'(R[[x]]))$ is also infinity.

Proof: For every positive integer n, it is enough to construct a complete subgraph of $\Gamma'(R[[x]])$ with n vertices. To this end, let n be an arbitrary positive integer and $a \in W * (R)$. Then, by Lemma 3.5, it is easy to see that the subgraph with vertex-set $\{a + x^{n+1}, \ldots, a + x^{2n}\}$ is a complete subgraph of $\Gamma'(R[[x]])$ which is isomorphic to K_n. So clique$(\Gamma'(R[[x]]))$ is infinity and this implies that $\chi(\Gamma'(R[[x]]))$ is infinity. We end this section with the following theorem. \

Theorem 3.8: The cozero-divisor graph $\Gamma'(R[[x]])$ is not planar.

Proof: In view of the proof of Theorem 3.7, k_5 is a subgraph of $\Gamma'(R[[x]])$. Thus, by Kuratowski's Theorem, $\Gamma'(R[[x]])$ is not planar.

COZERO-DIVISOR GRAPH OF R1XR2

Throughout this section, R_1 and R_2 are two commutative rings with non-zero identities. We will study the cozerodivisor graph of the direct product of R_1 and R_2. Note that an element (a,b)belongs to $W(R_1XR_2)$ if and only if $a \in W(R_1)$ or $b \in W(R_2)$. We begin this section with the following lemma.

Lemma 4.1: Suppose that $R = R_1 x \cdots x R_n$ is a direct product of finite commutative rings. If a_i is adjacent to b_i in $\Gamma'(R_i)$, for some $1 \leq i \leq n$, then every element in R with i-th component a_i is adjacent to all elements in R with i-th component b_i.

Proof: Suppose that a_i is adjacent to b_i in $\Gamma'(R_i)$ and assume on the contrary that the vertices $(a_1, \cdots, a_n)$ and $(b_1, \cdots, b_n)$ are not adjacent in $\Gamma'(R)$. Without loss of generality, suppose that $(a_1, \ldots, a_n) \in (b_1, \cdots, b_n)(R_1 x \cdots x R_n)$. Thus $(a_1, \cdots, a_n) = (b_1, \cdots, b_n)(r_1, \cdots, r_n)$, for some non-zero element $(r_1, \cdots, r_n) \in R$. Therefore $a_i = r_i b_i$ and hence a_i is not adjacent to b_i, which is a contradiction.

The following corollary follows immediately from Lemma 4.1.

Corollary 4.2: Suppose that (a,b), $(a',b') \in W*(R_1) x W*(R_2)$ such that they are not adjacent in $\Gamma'(R_1 x R_2)$. Then a is not adjacent to a' in $\Gamma'(R_1)$ and b is not adjacent to b' in $\Gamma'(R_2)$.

In the next lemma, we establish some relations between the adjacency in the graph $\Gamma'(R_1 x R_2)$ and adjacency in both graphs $\Gamma'(R_1)$ and $\Gamma'(R_2)$.

Lemma 4.3
- Let $a \in R_1$ and $b, b' \in R_2$. Then (a,b) is adjacent to (a,b') in $\Gamma'(R_1 x R_2)$ if and only if b is adjacent to b' in $\Gamma'(R_2)$.
- Let $b \in R_2$ and $a, a' \in R_1$. Then (a,b) is adjacent to (a,b') in $\Gamma'(R_1 x R_2)$ if and only if a is adjacent to a' in $\Gamma'(R_1)$.

Proof
1) Suppose that (a,b) is adjacent to (a,b'). Note that if at least one of the elements b or b' is zero or unit, then (a,b) is not adjacent to (a,b') in $\Gamma'(R_1 x R_2)$. Thus we can suppose that $b, b' \in W*(R_2)$. Now, if b is not adjacent to b', then $b \in b'R_2$ or $b' \in bR_2$. So without loss of generality, we may assume that $b = rb'$ for some non-zero element $r \in R_2$. Hence $(a,b) = (1,r)(a,b')$. This means that (a,b) and (a,b') are not adjacent in $\Gamma'(R_1 x R_2)$ which is impossible. Therefore b and b' are adjacent in $\Gamma'(R_2)$. Conversely, if b is adjacent to b', then by Lemma 4.1, we have that (a,b) is adjacent to (a,b').
2) The proof is similar to part 1).

The following propositions follow directly from Lemma 4.3.

Proposition 4.4: Assume that either $\Gamma'(R_1)$ or $\Gamma'(R_2)$ is not planar. Then $\Gamma'(R_1 \times R_2)$ is not planar.

Proof: Without loss of generality, suppose that $\Gamma'(R_1)$ is not planar. So, by Kuratowski's Theorem (cf. [8, p. 153]), it contains a subdivision of K_5 or $K_{3,3}$. Now, by Lemma 4.3 (ii), one can conclude that $\Gamma'(R_1 \times R_2)$ is

$$\text{clique}\left(\Gamma'\left(R_1 \times R_2\right)\right)$$

$$\geq \text{Max}\left\{\text{clique}\left(\Gamma'\left(R_1\right)\right), \text{clique}\left(\Gamma'\left(R_2\right)\right)\right\};$$ lowing inequalities:

$$\chi\left(\Gamma'\left(R_1 \times R_2\right)\right) \geq \text{Max}\left\{\chi\left(\Gamma'\left(R_1\right)\right), \chi\left(\Gamma'\left(R_2\right)\right)\right\}.$$

Remark 4.6: Suppose that $a \in R_1^*$ and $b \in R_2^*$. Then $(a,0)$ is adjacent to $(b,0)$.

In the following theorem, we invoke the previous lemmas to show that $\Gamma'(R_1 \times R_2)$ is a complete bipartite graph whenever R_1 and R_2 are fields.

Theorem 4.7: Assume that R_1 and R_2 are fields. Then $\Gamma'(R_1 \times R_2)$ is a complete bipartite graph.

Proof: Put $V_2 := \{(0,b) \mid b \in R_2^*\}$. Clearly $V_1 \cup V_2 = W^*(R_1 \times R_2)$. By Remark 4.6, every element in V_1 is adjacent to all elements of V_2 and vice versa. Also, it is easy to see that there is no adjacency between vertices in V_1 (or V_2). So $\Gamma'(R_1 \times R_2)$ is a complete bipartite graph.

Corollary 4.8: Let $\mathbb{F}$ be an arbitrary field. Then $\Gamma'(\mathbb{Z}_2 \times \mathbb{F})$ and $\Gamma'(\mathbb{F} \times \mathbb{Z}_2)$ are star graphs.

Remark 4.9: It is easy to see that (a,b) is adjacent to $(0,b')$ in $\Gamma'(R_1 \times R_2)$, for any $a \in R_1^*$, $b \in W(R_2)$ and $b' \in U(R_2)$. Similarly, (a,b) is adjacent to $(a',0)$ $\Gamma'(R_1 \times R_2)$, for any, $a \in W(R_1), b \in R_2^*$ and $a' \in U(R_1)$.

The following theorem is one of our main results in this section.

Theorem 4.10: The cozero-divisor graph $\Gamma'(R_1 \times R_2)$ is connected and $\text{diam}(\Gamma'(R[[x]])) \leq 3$

Proof: Suppose that (a,b) and (a',b') are arbitrary elements in $W^*(R_1 \times R_2)$. We have the following cases for consideration:

Case 1: $a, a' \in W(R_1)$. If $b = 0 = b'$, then consider the path $(a, 0) - (0, 1) - (a', 0)$. If $b \neq 0$ and $b' \neq 0$, then, by Remark 4.9, we have that $(a, 0) - (0, 1) - (a', b')$. Now, suppose that $b \neq 0$ and $b' = 0$. Then, in view of Remarks 4.6 and 4.9, one can obtain the path $(a, b) - (1, 0) - (0, 1) - (a', 0)$ in $\Gamma'(R_1 \times R_2)$. The similar result holds in the case that $b \neq 0$ and $b = 0$.

Case 2: $a \in W(R_1)$ and $a' \notin W(R_1)$. If $b \neq 0$, then, by Remark 4.9, whenever $b' = 0$, we have that $(a, b) = (a', b')$. Otherwise, $b' \neq 0$. Since $(a', b') \in W^*(R_1 + R_2)$ and $a' \notin W(R_1)$, we have $b' \in W(R_2)$. Now, if $(0, 1) \in (a', b')(R_1 \times R_2)$, then $1 \in b'R_2$. But $b'R_2 \subseteq W(R_2)$ and this implies that $1 \in W(R_2)$ which is not true. Hence $(0, 1) \notin (a', b')(R_1 \times R_2)$. Also, since $a' \neq 0$, it is easy to see that $(a', b') \notin (0, 1)(R_1 \times R_2)$. Thersefore, we have the path $(a, b) - (1, 0) - (0, 1) - (a', b')$. Also, if $b=0$, then by Remarks 4.6 and 4.9, we can consider the path $(a, 0) - (0, 1) - (a', b')$. The similar result holds if $a \notin W(R_1)$ and $a' \notin W(R_1)$.

Case 3: $a, a' \notin W(R_1)$. Then we have that $b, b' \notin W(R_2)$, and we can apply Case 1 on the second component of ordered pairs.

Now, in view of the above cases, it is easy to see that $\mathrm{diam}(\Gamma'(R_1 \times R_2)) \leq 3$.

In the next proposition, we provide a characterization of the complete cozero-divisor graph $\Gamma'(R_1 \times R_2)$.

Proposition 4.11: The graph $(\Gamma'(R_1 \times R_2))$ is complete if and only if $R_1 \times R_2$ is isomorphic to $\mathbb{Z}_2 \times \mathbb{Z}_2$.

Proof: If $|R_1| > 2$, then $(1, 0)$ is not adjacent to $(0, a)$, for some $1 \neq a \in R_1^*$. Similarly, if $|R_1| > 2$, then $\Gamma'(R_1 \times R_2)$ is not complete. So, if $\Gamma'(R_1 \times R_2)$ is complete, then $R_1 \times R_2 \cong \mathbb{Z}_2 \times \mathbb{Z}_2$. Also, clearly the graph $\Gamma'(\mathbb{Z}_2 \times \mathbb{Z}_2)$ is complete.

Corollary 4.12: If $R_1 \times R_2 \neq \mathbb{Z}_2 \times \mathbb{Z}_2$, then $\mathrm{diam}(\Gamma'(R_1 \times R_2)) = 2$ or 3.

In the following theorem, we study the girth of $\Gamma'(R_1 \times R_2)$. Note that we consider $\Gamma'(\mathbb{Z}_2)$ to be totally disconnected.

Theorem 4.13
- If at least one of the cozero-divisor graph $\Gamma'(R_1)$ or $\Gamma'(R_2)$ is not totally disconnected, then $g(\Gamma'(R_1 \times R_2)) = 3$.

- If $R_1 \neq \mathbb{Z}_2$ and $R_2 \neq \mathbb{Z}_2$, then $g(\Gamma'(R_1 \times R_2)) \leq 4$.
- If $R_1 \neq \mathbb{Z}_2$ and R_2 is a field, then $g(\Gamma'(R_1 \times R_2)) = \infty$.
- If $R_1 \neq \mathbb{Z}_2$ and R_2 is not a field, then $g(\Gamma'(R_1 \times R_2)) \leq 3, 4$ or ∞.

Proof

1) Without loss of generality, suppose that $a, b \in W^*(R_2)$ such that a is adjacent to b. Now, by Lemma 4.3 (i) and Remark 4.9, we have the cycle $(0,1) - (1,a) - (1,b) - (0,1)$ in $\Gamma'(R_1 \times R_2)$.
2) Let $a \in R_1^*$ and $b \in R_2^*$ such that a and b are not identity. Now, consider the cycle $(a,0) - (0,1) - (1,0) - (0,b) - (a,0)$ in $\Gamma'(R_1 \times R_2)$.
3) By Corollary 4.8, $\Gamma'(\mathbb{Z}_2 \times \mathbb{F})$ is a star graph and so $g(\Gamma'(\mathbb{Z}_2 \times \mathbb{F})) = \infty$.
4) First, assume that $|U(R_2)| > 1$. Let $a \in W^*(R_2)$ and $1 \neq b \in U(R_2)$. Now, by Remarks 4.6 and 4.9, we have the cycle $(0,1) - (1,0) - (0,b) - (1,a) - (0,1)$ in $\Gamma'(R_1 \times R_2)$. In the case that $|U(R_2)| = 1$, if $\Gamma'(R_2)$ is not totally disconnected, then by part 1), $g(\Gamma'(R_1 \times R_2)) = 3$. So, assume that there is no adjacency in $\Gamma'(R_2)$. In this situation, we first show that $\Gamma'(R_1 \times R_2)$ is a bipartite graph. To this end, set $v_1 := \{(0,a) \mid a \in R_2^*\}$ and $v_2 := \{(1,b) \mid b \in W(R_2)\}$. Clearly, $W^*(R_1 \times R_2) = V_1 \cup V_2$. Also, by Lemma 4.3, no two vertices in V_1 (or V_2) are adjacent. So $\Gamma'(R_1 \times R_2)$ is a bipartite graph, and thus $g(\Gamma'(R_1 \times R_2)) = 4$ or ∞. Now, by Remark 4.9, $(0,1)$ is adjacent to all vertices $(1, b)$ in $\Gamma'(R_1 \times R_2)$, where $b \in W^*(R_2)$, and also $(1,0)$ is adjacent to all vertices $(0,a)$ in $\Gamma'(R_1 \times R_2)$, where $a \in R_2^*$. Hence, if there exist an element $a \in WR_2^* \setminus \{1\}$ and $b \in W^*(R_2)$ such that $(0,a)$ is adjacent to $(1, b)$ in $\Gamma'(R_1 \times R_2)$s, then $g(\Gamma'(R_1 \times R_2)) = 4$. Otherwise, the girth of the cozero-divisor graph $\Gamma'(R_1 \times R_2)$ is infinity.

The following example presents a ring $R_1 \times R_2$ with $U(R_2)| = 1$ which satisfies parts 1) and 4) of Theorem 4.13. This shows that all cases in the proof of the last part of Theorem 4.13, can occur.

Example 4.14: Let $R_1 = \mathbb{Z}_2$ and $R_2 = \mathbb{Z}_2 \times \mathbb{Z}_2$. Then $|U(\mathbb{Z}_2 \times \mathbb{Z}_2)| = 1$ and by Proposition 4.11, $\Gamma'(\mathbb{Z}_2 \times \mathbb{Z}_2)$ is complete. Hence, by Theorem 4.13 (i), we have that $g(\Gamma'(\mathbb{Z}_2 \times (\mathbb{Z}_2 \times \mathbb{Z}_2))) = 3$.

ACKNOWLEDGMENTS

The authors are deeply grateful to the referee for careful reading of the manuscript and helpful suggestions.

REFERENCES

1. I. Beck, "Coloring of Commutative Rings," Journal of Algebra, Vol. 116, No. 1, 1988, pp. 208-226. doi: 10.1016/0021-8693(88)90202-5.
2. D. F. Anderson and P. S. Livingston, "The Zero-Divisor Graph of a Commutative Ring," Journal of Algebra, Vol. 217, No. 2, 1999, pp. 434-447. doi:10.1006/jabr.1998.7840.
3. M. Axtell, J. Coykendall and J. Stickles, "Zero-Divisor Graphs of Polynomials and Power Series over Commutative Rings," Communications in Algebra, Vol. 33, No. 6, 2005, pp. 2043-2050. doi: 10.1081/AGB-200063357.
4. T. G. Lucas, "The Diameter of a Zero Divisor Graph," Journal of Algebra, Vol. 301, No. 1, 2006, pp. 174-193. doi:10.1016/j.jalgebra.2006.01.019.
5. M. Axtell, J. Stickles and J. Warfel, "Zero-Divisor Graphs of Direct Products of Commutative Rings," Houston Journal of Mathematics, Vol. 32, No. 4, 2006, pp. 985-994.
6. M. Afkhami and K. Khashyarmanesh, "The Cozero-Divisor Graph of a Commutative Ring," Southeast Asian Bulletin of Mathematics, Vol. 35, No. 5, 2011, pp. 753- 762.
7. G. Chartrand and O. R. Oellermann, "Applied and Algorithmic Graph Theory," McGraw-Hill, Inc., New York, 1993.
8. J. A. Bondy and U. S. R. Murty, "Graph Theory with Applications," American Elsevier, New York, 1976.

CITATION

M. Afkham and K. Khashyarmanesh, "On the Cozero-Divisor Graphs of Commutative Rings," Applied Mathematics, Vol. 4 No. 7, 2013, pp. 979-985. doi: 10.4236/am.2013.47135.

The Continuous Analogy of Newton's Method for Solving a System of Linear Algebraic Equations

Tugal Zhanlav[1], Ochbadrakh Chuluunbaatar[1,2], and Gantumur Ankhbayar[1]
[1]School of Mathematics and Computer Science, National University of Mongolia, Ulan-Bator, Mongolia
[2]Joint Institute for Nuclear Research, Dubna, Moscow Region, Russia

ABSTRACT

We propose a continuous analogy of Newton's method with inner iteration for solving a system of linear algebraic equations. Implementation of inner iterations is carried out in two ways. The former is to fix the number of inner iterations in advance. The latter is to use the inexact Newton method for solution of the linear system of equations that arises at each stage of outer iterations. We give some new choices of iteration parameter and of forcing term, that ensure the convergence of iterations. The performance and efficiency of the proposed iteration is illustrated by numerical examples that represent a wide range of typical systems.

INTRODUCTION

We consider a system of linear algebraic equations

$$Ax = f. \tag{1}$$

For a numerical solving of Equation (1) we consider an iterative process:

$$Bv_n = -r_n, x^{(n+1)} = x^{(n)} + \tau_n v_n, n = 0, 1, \cdots$$

(2)

Of course, the quality of iterative process Equation (2) essentially depends on the choices of matrix B and of iteration parameter $\tau_n \neq 0$

Let H be a linear space of m-dimensional vectors. We will denote the Euclidean vector norm in H by $\|\cdot\|$, as well as the corresponding norm of matrices.

Theorem 1.1: Let $(Av_n, r_n) \neq 0$ Then a necessary and sufficient condition for the $\|r_n\|$ to be decreasing $\|r_{n+1}\| < \|r_n\|$ is that

$$\tau_n \in I_n = \begin{cases} \left(0, -\dfrac{2(Av_n, r_n)}{\|Av_n\|^2}\right), & \text{when } (Av_n, r_n) < 0, \\[4mm] \left(-\dfrac{2(Av_n, r_n)}{\|Av_n\|^2}, 0\right), & \text{when } (Av_n, r_n) > 0. \end{cases}$$

Proof: From Equation (2) we obtain

$$r_{n+1} = r_n + \tau_n Av_n.$$

Hence we have

$$\|r_{n+1}\|^2 = \|r_n\|^2 + \tau_n [\tau_n \| Av_n \|^2 + 2(Av_n, r_n)].$$

(3)

The assertion follows from (3).

The interval I_n we call τ-region of convergence of the iteration method (2). Thus we have to choose τ_n from this region. Moreover, it is desirable that the τ_n to be optimal in some sense. Further we will use well-known assertions to study the convergence of (2).

Theorem 1.2: [1]. Let S be an m × m matrix. Then the successive approximations

$$x^{(n+1)} = Sx^{(n)} + z, \ n = 0,1,2,\cdots \tag{4}$$

converge for each $z \in R^m$ and each $x^{(0)} \in R^m$ if and only if

$$\rho(S) < 1, \tag{5}$$

where $\rho(S)$ is a spectral radius of the matrix S.

It is easy to show that the iteration process (2) can be rewritten as (4) with iteration matrix

$$S = E - \tau_n B^{-1} A \text{ and } z = \tau_n B^{-1} f. \tag{6}$$

Here E is an m × m unit matrix.

Theorem 1.3: [2]. The iteration process (2) with parameter τ_n given by

$$\tau_n^* = -\frac{(Av_n, r_n)}{\|Av_n\|^2} \tag{7}$$

converges to x^* for any $x^{(0)} \in H$. The following a posteriori estimate holds true:

$$\|r_n\| = q_{n-1} q_{n-2} \cdots q_0 \|r_0\|, \tag{8}$$

where

$$q_i = \sqrt{1 - \frac{(Av_i, r_i)^2}{\|Av_i\|^2 \|r_i\|^2}} < 1, \ i = 0,1,\cdots,n-1.$$

We call the nonzero value of τ_n^* defined by (10) the optimal one in the sense that it yields the minimum value of functional $\|r_{n+1}\|$.

THE CONTINUOUS ANALOGY OF NEWTON'S METHOD

The continuous analogy of Newton's method is also applicable to (1) and leads to [2]

$$Av_n = -r_n,$$ (9a)

$$x^{(n+1)} = x^{(n)} + \tau_n v_n, n = 0,1,\cdots$$ (9b)

It should be mentioned that not only the convergence, but also the convergence rate of iteration (9) depends on the choice of the parameter τ_n. We have the following:

Lemma 2.1: The sufficient condition for $\|r_n\|$ to be decreasing is $0 < \tau_n < 2, n = 0,1,\cdots$

Lemma 2.2: Suppose that $1 < \tau_n < 2$ for all $n \geq 0$. Then the iteration process (2) gives two-sided approximations to x^*, i.e.,

$$x^{(1)} < \cdots < x^{(2m+1)} < x^* < x^{(2m)} < \cdots < x^{(0)} \text{ if } x^* < x^{(0)}$$

or

$$x^{(0)} < \cdots < x^{(2m)} < x^* < x^{(2m+1)} < \cdots < x^{(1)} \text{ if } x^{(0)} < x^*.$$

The proofs are immediately followed from the equalities

$$r_{n+1} = (1 - \tau_n) r_n, \ x^{(n+1)} - x^* = (1 - \tau_n)(x^{(n)} - x^*).$$

At each step of iterations one can solve the system (9a) by means of some iterative methods. We call this inner iteration. We consider the following decomposition of A:

$$A = A_1 + A_2,$$

in which the matrix A_1 is simple and invertible. For finding the correction v_n we use inner iteration

$$A_1 v_n^{(e)} = -r_n - A_2 v_n^{(e-1)}, \ v_n^{-1} \equiv 0, \ e = 0, 1, \cdots \tag{10}$$

Theorem 2.3: Suppose that

$$\|C\| < 1, \ C = A_2 A_1^{-1}. \tag{11}$$

Then the inner iteration (10) converges and holds the following estimate

$$\left\| v_n - v_n^{(k)} \right\| \leq \left\| A_1^{-1} \right\| \frac{\|C\|^{k+1}}{1 - \|C\|} \left\| r_n \right\|, k = 0, 1, \cdots \tag{12}$$

Proof: The linear system (10) can be rewritten as

$$(E + C) A_1 v_n = -r_n. \tag{13}$$

By assumption (11) there exists $(E + C)^{-1}$ and the following series representation is valid

$$(E + C)^{-1} = \sum_{j=0}^{\infty} (-1)^j C^j. \tag{14}$$

Then from (13) it follows that

$$v_n = -A_1^{-1} \left(\sum_{j=0}^{\infty} (-1)^j C^j \right) r_n. \tag{15}$$

In a similar way, from (10) we have

$$v_n^{(k)} = -A_1^{-1} \left(\sum_{j=0}^{k} (-1)^j C^j \right) r_n. \tag{16}$$

From (15) and (16) immediately follows (14).

The estimate (12) means that the inner iteration (10) converges under condition (11). In real computations we have to terminate the inner iteration before convergence. We will restrict the number of inner iteration by k. Then the whole iteration process looks like:

$$A_1 v_n^{(l)} = -r_n - A_2 v_n^{(l-1)}, \ v_n^{-1} \equiv 0, l = 0, 1, \cdots, k,$$

(17a)

$$x^{(n+1)} = x^{(n)} + \tau_n v_n^{(k)}, \ n = 0, 1 \cdots$$

(17b)

We now formulate the convergence theorems for these methods.

Theorem 2.4: Suppose that the condition (11) is satisfied and the iteration parameter τ_n is given by

$$\tau_n = -\frac{\left(A v_n^{(k)}, r_n \right)}{\left\| A v_n^{(k)} \right\|^2}.$$

(18)

Then the iteration process (17) converges for any $k = 0, 1, \cdots$ and for arbitrary chosen starting $x^{(0)}$.

Proof: We rewrite the iteration process (17) as (2) with

$$B = \left(\sum_{j=0}^{k} (-1)^j C^j \right)^{-1} A_1.$$

By Theorem 1.3, such a process converges if we choose τ_n by (7), in which the v_n is replaced by $v_n^{(k)}$.

Obviously the number k may be different for each n. From (12) it is evident, that it suffices to restrict the number of inner iteration only by

k = 0 or 1, when the residual norm $\left\| r_n \right\|$ is small enough.

Theorem 2.5: Suppose that the condition (11) is satisfied. Then the iteration process (17) with $\tau_n \in (0,1)$ converges for any $k = 0, 1, \cdots$ and for arbitrary chosen starting $x^{(0)}$. The following inequality holds:

$$\left\| r_{n+1} \right\| \leq \theta_n \left\| r_n \right\|, \ 0 < \theta_n < 1.$$

(19)

Proof: From (17) we get

$$r_{n+1} = r_n + \tau_n A v_n^{(k)}.$$

Using (11) and (16) we rewrite the last expression as

$$r_{n+1} = \left((1-\tau_n) E + \tau_n (-1)^{k+1} C^{k+1} \right) r_n. \tag{20}$$

If $0 < \tau_n < 1$, then from (20) we obtain

$$\|r_{n+1}\| \le \theta_n \|r_n\|, \quad \theta_n = 1 - \tau_n \left(1 - \|C\|^{k+1} \right).$$

According to (11), we have $0 < \theta_n < 1$. The convergence of (17) follows from (19).

Corollary 2.6: Let the condition (11) fulfill, and τ_n is given by [3]

$$\tau_n = \min \left\{ \frac{\|r_{n-1}\|}{\|r_n\|} \tau_{n-1}; 1 \right\}, \quad n = 1, 2, \cdots, \quad \tau_0 \approx 0.1.$$

Then the iteration (17) converges.

Remark 2.7: In proofs of Theorems 2.3-2.5 the condition (11) is essentially used. In particular, it may be fulfilled if the matrix A of the system (1) is a strongly diagonally dominant and A_1 is chosen as

$$A_1 = \operatorname{diag} \{ a_{11}, \cdots, a_{mm} \}.$$

Theorem 2.8: Suppose that the condition (11) is satisfied and that the eigenvalues of matrix $D = A_1^{-1} A_2$ are real. Then the spectral radius of iteration matrices of the iterations (17) is decreasing with respect to k when $\tau_n \to 1$.

Proof: The iterations (17) can be rewritten as (4) with iteration matrices

$$S_k = (1-\tau_n) E + \tau_n (-1)^{k+1} D^{k+1}. \tag{21}$$

We denote by $\lambda(D)$ the eigenvalues of D. If $\tau_n = 1$, then

$$S_k = (-1)^{k+1} D^{k+1}.$$

Therefore, we have

$$\left|\lambda\left(S_k\right)\right| = \left|\lambda\left(D\right)\right|^{k+1}.$$

By virtue of (11) we obtain $|\lambda(C)| \leq \| C \| < 1$. Since the spectrums of matrices C and D coincide, we also have

$$\left|\lambda\left(D\right)\right| < 1.$$

By definition we get

$$\rho\left(S_k\right) = \max\left|\lambda\left(S_k\right)\right| = \left(\max\left|\lambda\left(D\right)\right|\right)^{k+1}$$

$$= \left(\max\left|\lambda\left(D\right)\right|\right)^{k} \max\left|\lambda\left(D\right)\right|$$

$$< \max\left|\lambda\left(D\right)\right|^{k} = \rho\left(S_{k-1}\right),$$

which is valid for all $k = 1, 2, \cdots$ Thus we have

$$\rho\left(S_k\right) < \rho\left(S_{k-1}\right) < \cdots < \rho\left(S_1\right) < \rho\left(S_0\right) < 1. \tag{22}$$

Obviously, from (21) it is clear that S_k is a continuous function of τ_n.

Therefore, (22) is valid for $\tau_n \to 1$.   The iteration (17) with a few small k represents a special interest from a computational view point. Moreover, it is worth to stay at (17) with k = 0 in detail. The iteration (17) with k = 0 and $A_1 = diag\{a_{11}, \cdots, a_{mm}\}$ leads to the well-known Jacobi iteration with relaxation parameter $\omega = \tau_n$. It is also known that [1] the Jacobi method with optimal relaxation parameter

$$\omega_{opt} = \frac{2}{2 - \lambda_{\min} - \lambda_{\max}} \tag{23}$$

converges under the assumption that the Jacobi matrix $B = -A_1^{-1}A_2$ has real eigenvalues and spectral radius less than one. Here by $\lambda_{\min}$ and $\lambda_{\max}$ denoted are the smallest and largest eigenvalues of B. Fortunately,

we can prove the convergence of Jacobi method with relation parameter under a mild condition than the above mentioned assumption.
Namely, we have.

Corollary 2.9: Suppose that the condition (11) is satisfied. Then the
Jacobi method with relaxation parameter:

$$\omega_n = -\frac{\left(Av_n^{(0)}, r_n\right)}{\left\|Av_n^{(0)}\right\|} = \frac{\left((E+C)r_n, r_n\right)}{\left\|(E+C)r_n\right\|^2} \tag{24}$$

converges for any starting.

From (24) it is clear that the relaxation parameter changes depending
on n. Therefore the iteration (17) with k = 0, and $A_1 = diag\{a_{11}, \cdots, a_{mm}\}$
and with ω_n given by (24) we call nonstationary Jacobi method with
optimal relaxation parameters. The iteration (17) with k = 0 and with
a lower triangular matrix A_1 leads to GaussSeidel with one parameter.

The Formula (18) can be rewritten as:

$$\tau_n = \frac{\left(AB^{-1}r_n, r_n\right)}{\left\|AB^{-1}r_n\right\|^2}, \tag{25}$$

where $AB^{-1} = E - (-1)^{k+1}C^{k+1}$ From this it is clear that $\tau_n \to 1$ as $k \to \infty$.
This means that whole iteration (17) with the parameter given by (18)
converges quadratically in the limit.

Since $\tau_n \to 1$ as $k \to \infty$, then τ-region of convergence for iterations (17)
leads to

$$I_n^k = \left(0; -\frac{2\left(Av_n^{(k)}, r_n\right)}{\left\|Av_n^{(k)}\right\|^2}\right) \to (0; 2).$$

The number of outer iteration n depends on the number of inner iteration k, i.e., $n = n_k$. In general, n is a decreasing function of k, i.e., $n_{k+1} < n_k$

On the other hand, the iteration (17) can be considered as a defect correction iteration [1]

$$x^{(n+1)} = x^{(n)} - \tau_n A_{approx}^{-1} r_n, \left(B^{-1} = A_{approx}^{-1} \right), \tag{26}$$

where A_{approx}^{-1} defined by

$$A_{approx}^{-1} = A_1^{-1} \left(E - C + C^2 + \cdots + (-1)^k C^k \right). \tag{27}$$

A_{approx}^{-1} is a reasonable approximation to A^{-1} since

$$\left\| A^{-1} - A_{approx}^{-1} \right\| \le \frac{\left\| A_1^{-1} \right\| \left\| C \right\|^{k+1}}{1 - \left\| C \right\|} \to 0 \tag{28}$$

for large k. The choice of parameter τ_n given by (25) allows us to decrease the residual norm from iteration to iteration. By this reason we call (26) as a minimal defect (or residual) correction iteration.

From (26) it follows that

$$r_{n+1} = \left(E - \tau_n A A_{approx}^{-1} \right) r_n. \tag{29}$$

From (29) and $A = A_1 + A_2 = (E + C)A_1, C = A_2 A_1^{-1}$, it follows that

$$E - \tau_n A A_{approx}^{-1} = (1 - \tau_n) E + \tau_n (-1)^{k+1} C^{k+1}.$$

Therefore we have

$$\lambda \left(E - \tau_n A A_{approx}^{-1} \right) = 1 - \tau_n + \tau_n (-1)^{k+1} \left[\lambda(C) \right]^{k+1}.$$

This means that the spectral radius of matrix $E - \tau_n A A_{approx}^{-1}$ depends on the number of inner iteration k, i.e., $\rho = \rho(k)$. Therefore we have $\rho(k) \to 0$ as $k \to \infty$, because $\tau_n \to 1$ as $k \to \infty$ and from (11)

$$|\lambda(C)| \le \|C\| < 1.$$

Above we considered the cases, where the inner iteration number k is fixed at each outer iteration. It is desirable that the termination criterion for the inner iteration must be chosen carefully to preserve the superlinear convergence rate of the outer iteration. We stay at this problem in more detail. When $\tau_n = 1$, the iteration (17) is, indeed, inexact Newton (IN) method for (1). Therefore, iteration (17) with parameter τ_n given by formula (18) we call inexact damped Newton method (IDN).

According to IN method [4, 5], we must choose $\eta_n \in [0,1)$ and continue the inner iteration until satisfy the condition

$$\left\| r_n^{(e)} \right\| \le \eta_{n-1} \left\| r_n \right\|, \quad r_n^e = A v_n^{(e)} + r_n, \quad n = 1, 2, \cdots \tag{30}$$

Thus (30) is a stopping criterion for inner iteration (17a). There are several choices of forcing term η_n in inexact Newton method [4-6]. For examples, in [6] two choices of η_n were proposed:

$$\eta_n = \begin{cases} 1 - \eta_{n-1}\alpha_n, & \text{when } \eta_{n-1}\alpha_n < 1, \\ -\dfrac{1 - \eta_{n-1}\alpha_n}{\alpha_n}, & \text{when } \eta_{n-1}\alpha_n \ge 1, \end{cases} \tag{31}$$

with

$$\alpha_n = \frac{\|r_{n-1}\|}{\|r_n\|},$$

and

$$\eta_n = |1 - \tau_n|, \quad n = 0, 1, \cdots \tag{32}$$

Since we have formula (18) for τ_n, one can use the second choice (14) with no additional calculations. We can also use formula for $\tau_n = \dfrac{2}{1 + \sqrt{1 + \|r_n\|}}$ [7]. Therefore, according to (32), we get

$$\eta_n = \frac{\sqrt{1 + \|r_n\|} - 1}{\sqrt{1 + \|r_n\|} + 1}. \tag{33}$$

NUMERICAL RESULTS

The quality of the proposed iteration was checked up for numerous examples. We express the matrix A as

$$A = D + A_L + A_U,$$

where $D = diag\{a_{11}, \cdots, a_{mm}\}$, and A_L, A_U are lower and upper triangular matrices, respectively. All examples are calculated with an accuracy

$$\| Ax_n - f \| < 10^{-7}.$$

The numerical calculations are performed on Acer, CPU 1.8 GHz, 1 GB RAM, and using a software MATLAB R2007a for the Windows XP system.

Example 1: We consider a system of Equation (1) with matrix A and f given by

$$A = \begin{pmatrix} 2 & 1 & & & \\ 1 & 4 & 1 & & \\ & \ddots & \ddots & \ddots & \\ & & 1 & 4 & 1 \\ & & & 1 & 2 \end{pmatrix}_{m \times m}, f = 6 \times \begin{pmatrix} 0.5 \\ 1.0 \\ \vdots \\ 1.0 \\ 0.5 \end{pmatrix}_{m},$$

which was solved by the proposed iteration method (17), (18). For comparison, it was solved by Jacobi and successive over relaxation (SOR) iterations with a parameter ω_{opt}.

Example 2:

$$A = \begin{pmatrix} 1.1161 & 0.1254 & 0.1397 & 0.1490 \\ 0.1582 & 1.1675 & 0.1768 & 0.1871 \\ 0.1968 & 0.2071 & 1.2168 & 0.2271 \\ 0.2368 & 0.2471 & 0.2568 & 1.2671 \end{pmatrix}, f = \begin{pmatrix} 1.5471 \\ 1.6471 \\ 1.7471 \\ 1.8471 \end{pmatrix}.$$

Example 3:

$$A = \begin{pmatrix} 0.60 & -0.16 & -0.12 & -0.07 & -0.03 \\ -0.16 & 0.74 & -0.31 & -0.19 & -0.07 \\ -0.12 & -0.31 & 0.66 & -0.31 & -0.12 \\ -0.07 & -0.19 & -0.31 & 0.74 & -0.16 \\ -0.03 & -0.07 & -0.12 & -0.16 & 0.96 \end{pmatrix}, f = \begin{pmatrix} 1.0 \\ 0.1 \\ 0.1 \\ 0.1 \\ 1.0 \end{pmatrix}.$$

Example 4:

$$A = \begin{pmatrix} T_N + 2I_N & -I_N & & & \\ -I_N & T_N + 2I_N & -I_N & & \\ & \ddots & \ddots & \ddots & \\ & & -I_N & T_N + 2I_N & -I_N \\ & & & -I_N & T_N + 2I_N \end{pmatrix}_{N^2 \times N^2},$$

$$f = h^2 \begin{pmatrix} 1 \\ 1 \\ \vdots \\ 1 \\ 1 \end{pmatrix}_{N^2},$$

where $h = 1/(N+1)$ and

$$T_N = \begin{pmatrix} 2 & -1 & & & \\ -1 & 2 & -1 & & \\ & \ddots & \ddots & \ddots & \\ & & -1 & 2 & -1 \\ & & & -1 & 2 \end{pmatrix}_{N \times N}, I_N = \begin{pmatrix} 1 & & & & \\ & 1 & & & \\ & & \ddots & & \\ & & & 1 & \\ & & & & 1 \end{pmatrix}_{N \times N}.$$

Such a system arises in discretization of two dimensional Poisson equation [8]:

$$\Delta u(x,y) = -1, \quad D = \{(x,y) \mid 0 \le x, y \le 1\}, \quad u = 0 \text{ on } \partial D.$$

The exact value of $u(1/2, 1/2)$ is [9]

$$u\left(\frac{1}{2}, \frac{1}{2}\right)$$

$$= \frac{16}{\pi^4} \sum_{\mu,\nu=0}^{\infty} \frac{(-1)^{\mu+\nu}}{(1+2\mu)(1+2\nu)\left((1+2\mu)^2 + (1+2\nu)^2\right)} \tag{34}$$

The properties of matrix for examples 1-4 are shown in Table 1. From this we see that the considered examples represent a wide range of typical systems.

The numbers of Jacobi and SOR iterations versus the dimension m of example 1 are presented in Table 2. Here k is the number of inner iteration in (17) (18). From this example we see that the proposed iteration (17), (18) can compete with SOR iteration with optimal relaxation parameter and seem to be superior to the Jacobi iteration. The behavior of the iteration parameter τ_n given by (18) at m = 10 is explained in Table 3. From this we see that the iteration parameter τ_n tends to 1 as k increases.

The number n of outer iteration (17), (18) with fixed number k of inner iterations, and CPU time for examples 2 and 3 are shown in Table 4. From this we see that they are in example 2 considerable less than example 3. This explained by reason that matrix of this system has a strictly diagonally dominant. The number n of outer iteration, the total number k of inner iteration when forcing terms η_n was chosen by formulas (32) and (33), and CPU time are displayed in Table 5. Here it is observed similar situations as in the previous case.

Monotonic convergence of the calculated values $u^h(1/2,1/2)$ to exact value (34) versus the dimension N of the example 4 is shown in Table 6. The number n of outer iteration with fixed and unfixed number k of inner iteration and CPU time versus the dimension N of the example 4 are presented in Tables 7 and 8, respectively.

Table 1: The properties of matrix for examples 1–4			
Example	Matrix		
	Symmetric	Diagonally dominant	Sparse
1	+	+	+
2	-	+	-
3	+	-	-
4	+	-	+

Table 2: The numbers of Jacobi and SOR iterations versus the dimension m of example 1. Here k is the number of inner iteration in (17), (18)

Iter.	A1=D				Jacobi method	A1=AL+D				SOR
m/k	0	1	2	3		0	1	2	3	ω_{opt}
10	16	10	9	6	30	12	7	5	4	12
100	18	9	8	5	32	14	7	5	4	17
1000	17	9	8	5	34	14	7	5	4	18

Table 3: The behavior of the iteration parameter τ_n given by (18) for example 1. Here n and k are the numbers of outer and inner iterations in (17), respectively

n/k	1	2	3
1	1.031939	0.994921	1.000760
2	0.998348	1.001490	0.998875
3	0.988243	0.999830	1.000464
4	0.989728	1.011261	0.999378
5	1.014757	1.004592	
6	1.024398		
7	1.053859		

Table 4: The number n of outer iteration with fixed number k of inner iteration, and CPU time CT for examples 2 and 3

A1 K=0		Example 2			Example 3		
		K=1	K=2	K=0	K=1	K=2	
D	n	39	17	13	196	89	56
	CT	1.5e-3	9.2e-4	8.6e-4	8.6e-4	6.9e-3	3.1e-3
AL	n	14	5	13	92	58	41
	CT	6.7e-3	4.9e-4	4.5e-4	3.3e-4	2.7e-3	2.3e-3
SOR	n		8			35	
	CT		1.3e-3			5.1e-3	

Table 5: The number n of outer iteration, the total number k of inner iteration when forcing terms η_n was chosen by formulas (32) and (33), and CPU time CT

A1				Example 2	Example 3
D	(33)	CT		1.61e-3	1.07e-2
		n(k)		4(12)	5(188)
	(32)	CT		1.72e-3	1.08e-2
		n(k)		5(11)	5(135)
AL+D	(33)	CT		1.10e-3	7.53e-3
		n(k)		4(5)	5(94)
	(32)	CT		1.11e-3	7.42e-3
		n(k)		4(4)	4(92)

CONCLUSIONS

Our method with inner iteration is quadratically convergent and therefore it can compete with other iterations such as SOR with an optimal relaxation parameter for a strictly diagonally dominant system. Moreover, our method is also applicable not only for the system with a strictly diagonal dominant matrix, but also for system, the matrix of which is not Hermitian and positive definite.

This work was partially sponsored by foundation for science and technology of Ministry of Education, Culture, and Science (Mongolia).

The Continuous Analogy of Newton's Method for Solving a System

Table 6: A comparison of the calculated values u(1/2, 1/2) and an exact value (34) versus the dimension N of the example 4

N + 1	u(1/2, 1/2)
4	0.070312
8	0.072783
16	0.073446
32	0.073615
Exact	0.073671

Table 7: The number n of outer iteration with fixed number k of inner iteration, and CPU time CT versus the dimension N of the example 4

N+1 k=0		A1=D		A1=AL+D			Tridiagonal				SOR
		K=1	K=2	K=0	K=1	K=2	K=0	K=1	K=2		
4	n	64	20	21	32	15	11	39	16	12	11
	CT	2.5e-3	1.3e-3	1.4e-3	1.3e-3	8.8e-4	8.1e-4	1.2e-3	8.3e-4	7.6e-4	1.3e-3
8	n	267	60	87	124	60	40	149	40	47	23
	CT	1.4e-2	9.6e-3	1.2e-2	6.7e-3	3.9e-3	4.2e-3	7.9e-3	2.8e-3	3.6e-3	6.1e-3
16	n	1010	159	328	476	214	137	546	99	183	46
	CT	5.9e-1	1.4e-1	3.0e-1	2.8e-1	1.7e-1	1.3e-1	2.9e-1	1.0e-1	1.8e-1	1.0e-1

Table 8: The number n of outer iteration with unfixed number k of inner iteration, the itration parameter τ_n, and CPU time CT versus the dimension N of the example 4

N+1	n	A1=D			A1=AL+D			Tridiagonal		
		k		CT	k	τ_n	CT	k	τ_n	CT
4	1	9	1.0304	3.58e–3	5	1.0205	2.42e–3	5	1.0267	2.22e–3
	2	4	1.0087		3	0.9845		2	1.0207	
	3	18	1.0013		10	1.0006		8	1.0037	
8	1	44	1.0235	3.37e–3	23	1.0202	1.67e–3	23	1.0213	1.77e–2
	2	35	1.0108		14	1.0118		18	1.0102	
	3	70	1.0030		34	1.0032		35	1.0034	
16	1	211	1.0122	6.07e–0	106	1.0120	3.07e–0	106	1.0122	3.10e–0
	2	185	1.0072		89	1.0075		93	1.0072	
	3	195	1.0167		99	1.0158		98	1.0165	

O. Chuluunbaatar acknowledges a financial support from the RFBR Grant No. 11-01-00523, and the theme 09-6-1060-2005/2013 "Mathematical support of experimental and theoretical studies conducted by JINR".

REFERENCES

1. R. Kress, "Numerical Analysis," Springer-Verlag, Berlin, 1998. doi:10.1007/978-1-4612-0599-9
2. T. Zhanlav, "On the Iteration Method with Minimal Defect for Solving a System of Linear Algebraic Equations," Scientific Transaction, No. 8, 2001, pp. 59-64.
3. T. Zhanlav and I. V. Puzynin, "The Convergence of Iterations Based on a Continuous Analogy Newton's Method," Journal of Computational Mathematics and Mathematical Physics, Vol. 32, No. 6, 1992, pp. 729-737.
4. R. S. Dembo, S. C. Eisenstat and T. Steihaug, "Inexact Newton Methods," SIAM Journal on Numerical Analysis, Vol. 19, No. 2, 1982, pp. 400-408. doi:10.1137/0719025
5. H. B. An, Z. Y. Mo and X. P. Liu, "A Choice of Forcing Terms in Inexact Newton Method," Journal of Computational and Applied Mathematics, Vol. 200, No. 1, 2007, pp. 47-60.doi:10.1016/j.cam.2005.12.030
6. T. Zhanlav, O. Chuluunbaatar and G. Ankhbayar, "Relationship between the Inexact Newton Method and the Continuous Analogy of Newton Method," Revue D'Analyse Numerique et de Theorie de L'Approximation, Vol. 40, No. 2, 2011, pp. 182-189.

7. T. Zhanlav and O. Chuluunbaatar, "The Local and Global Convergence of the Continuous Analogy of Newton's Method," Bulletin of PFUR, Series Mathematics, Information Sciences, Physics, No. 1, 2012, pp. 34-43.
8. J. W. Demmel, "Applied Numerical Linear Algebra," SIAM, Philadelphia, 1997, pp. 265-360. doi:10.1137/1.9781611971446
9. W. Hackbusch, "Elliptic Differential Equations: Theory and Numerical Treatment," Springer Series in Computational Mathematics, Vol. 18, Springer, Berlin, 1992.

CITATION

T. Zhanlav, O. Chuluunbaatar and G. Ankhbayar, "The Continuous Analogy of Newton's Method for Solving a System of Linear Algebraic Equations," Applied Mathematics, Vol. 4 No. 1A, 2013, pp. 210-216. doi: 10.4236/am.2013.41A032.

Discrete Differential Geometry of n-Simplices and Protein Structure Analysis

Naoto Morikawa
Genocript, Zama, Japan

ABSTRACT

This paper proposes a novel discrete differential geometry of n-simplices. It was originally developed for protein structure analysis. Unlike previous works, we consider connection between space-filling n-simplices. Using cones of an integer lattice, we introduce tangent bundle-like structure on a collection of n-simplices naturally. We have applied the mathematical framework to analysis of protein structures. In this paper, we propose a simple encoding method which translates the conformation of a protein backbone into a 16-valued sequence.

INTRODUCTION

This paper proposes a novel discrete differential geometry of n-simplices, which is originally developed for protein structure analysis [1] [2]. Discrete differential geometry is the study of discrete equivalents of the geometric notions and methods of classical differential geometry [3] [4]. It mainly deals with polygonal curves and polyhedral surfaces whose properties are analogous to continuous counterparts, where the smooth theory is established as limit of the discrete theory.

On the other hand, we consider connection between space-filling n-simplices. We define gradient of n-simplices and obtain a flow of n-simplices by piling up n-cubes diagonally. Second derivative along a trajectory is given as a binary-valued sequence for any n (>1). As a result, we could encode the shape of n-dimensional objects if we approximate them by sweeping the occupied area with a trajectory of n-simplices.

Proteins are a sequence of amino acids linked by peptide bonds and fold into a unique three-dimensional structure in nature. Protein backbone structure is usually studied via manually-curated hierarchical classification [5] [6] but there also exist studies on differential geometric approach for protein structure analysis [7] - [11]. As for discrete differential geometry of protein backbones, proteins are usually represented as a polygonal chain, where curvature and torsion are defined at each vertex [7].

In our method, protein backbone structures are approximated by a trajectory of 3-simplices (tetrahedrons). Particularly we consider second derivative along a trajectory to encode local protein structures. Our method performs comparably with more sophisticated but more time-consuming methods which are specifically designed for protein structure analysis [12] [13]. In the following, we first describe the discrete differential geometry of n-simplices. Then, we apply the mathematical framework to analysis of protein structures and propose a simple encoding method which translates the conformation of a protein backbone into a 16-valued sequence.

DISCRETE DIFFERENTIAL GEOMETRY OF n-SIMPLICES

Basic Ideas

Recall that an n-simplex is an n-dimensional polytope which is the convex hull of its n + 1 vertices. As an introduction, we would consider the case of n = 2 before we give the definitions in the general case. In the case of n = 2, we obtain a flow of 2-simplices (triangles) by piling up unit cubes in the three-dimensional Euclidean space $\mathbb{R}^3$ as shown in Figure 1(a).

First, cubes are piled up in the direction of (-1,-1,-1), where three upper faces of each unit cube are divided into two triangles by a diagonal line. Then, the diagonal lines on the faces of the cubes form a drawing on the surface of the "peaks and valleys" of cubes. By projecting the drawing onto a hyperplane that is perpendicular to (1, 1, 1), a flow of triangles would be obtained. For example, the grey "slant" triangles on the surface specify the closed trajectory of the grey "flat" triangles on the hyperplane in Figure 1(a).

Differential Structure

Because of convenience, we use monomials to represent coordinates of points. That is, point $(l_1, l_2, \cdots, l_n) \in \mathbb{R}^n$ is denoted by monomial $x_1^{l_1} x_2^{l_2} ... x_n^{l_n}$ of n indeterminates for integer n (n > 1).

First of all, we give the definition of "slant" and "flat" n-simplices. Let's consider n-cube in the n-dimensional Euclidean space $\mathbb{R}^n$. Note that the facets of n-cubes are n-1 -dimensional unit cubes. To obtain "slant" n-simplices, we divide each of the n facets which contain origin (0, 0 ,..., 0) into (n-1) (n-2) n-1 -simplices along diagonal as follows.

Definition 1: For any integer n > 1, n-dimensional standard lattice L_n is the collection of all integer points of $\mathbb{R}^n$, i.e.,

$$L_n = \left\{ x_1^{l_1} x_2^{l_2} \cdots x_n^{l_n} \middle| l_i \in \mathbb{Z} \text{ for } 1 \le i \le n \right\}.$$

Definition 2: For any integer n > 1, the collection S_n of all slant n-simplices is defined by

$$S_n = \left\{ a\left[x_{\rho(1)} \cdots x_{\rho(n-1)} \right] \middle| a \in L_n, \rho \in Sym_n \right\},$$

Where Sym_n is the n-th symmetric group and $a[x_{\rho(1)}...x_{\rho(n-1)}]$ denotes the convex hull of n points

$$a_0 = a, a_1 = aX_{\rho(1)}, \cdots, a_{n-1} = aX_{\rho(1)}X_{\rho(2)}\cdots X_{\rho(n-1)} \in \mathbb{R}^n$$

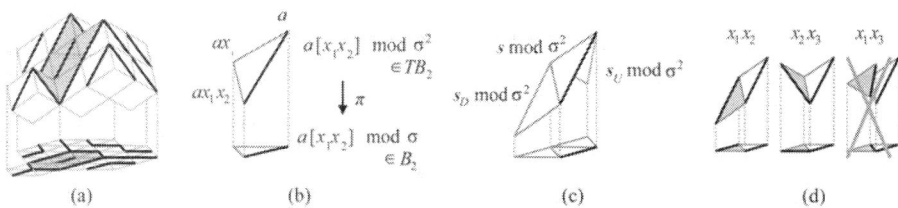

(a)   (b)   (c)   (d)

Figure 1: Discrete differential geometry of 2-simplices: (a) "Peaks and valleys" of cubes; (b) Tangent bundle-like structure; (c) Local trajectory; (d) Smoothness condition.

i.e.,

$$a\left[X_{\rho(1)}\cdots X_{\rho(n-1)}\right] = \left\{\prod_{0\le i<n} a_i^{\lambda_i} \,\middle|\, \lambda_i \in \mathbb{R} \ (0\le i<n) \text{ s.t. } \lambda_i \ge 0 \text{ and } \sum_{0\le i<n}\lambda_i = 1\right\}.$$

Definition 3: For any integer n > 1, the collection B_n of all flat n-simplices is defined as the quotient of S_n by "shift operator" σ on S_n, i.e,

$$B_n = S_n/\sigma,$$

Where

$$\sigma\left(a\left[X_{\rho(1)}\cdots X_{\rho(n-1)}\right]\right) = aX_{\rho(1)}\left[X_{\rho(2)}\cdots X_{\rho(n)}\right].$$

Definition 4: Tangent bundle-like structure TB_n is defined on B_n as the quotient of S_n by σ^n, i.e.,

$$\begin{cases} TB_n = S_n/\sigma^n, \\ \pi: TB_n \to B_n, \ \pi\left(s\bmod\sigma^n\right) = s\bmod\sigma. \end{cases}$$

For example, triangle $a[x_1x_2]$ specifies element $a[x_1x_2]\bmod\sigma^2$ of TB_n over element $a[x_1x_2]\bmod\sigma$ of B_n (Figure 1(b)).

Definition 5: Gradient Ds of $s = a[x_{\rho(1)}...x_{\rho(n-1)}] \in S_n$ is defined as a monomial of degree $n - 1$, i.e.,

$$Ds = x_{\rho(1)} x_{\rho(2)} \cdots x_{\rho(n-1)}.$$

For simplicity, we occasionally denote $x_{\rho(1)}...x_{\rho(n-1)}$ by $e / x_{\rho(n)}$, where $e = x_1 x_2 ... x_n$. That is $Ds = e / x_{\rho(n)}$.

Note that we could identify TB_n with $B_n \times \{e / x_1, e / x_2, ..., e / x_n\}$ by one-to-one correspondence

$$s \bmod \sigma^n \sim (s \bmod \sigma, Ds).$$

A gradient over a flat n-simplex specifies a local trajectory at the flat n-simplex as follows.

Definition 6: The local trajectory specified by $s \bmod \sigma^n \in TB_n$, where $s = a[x_{\rho(1)}...x_{\rho(n-1)}] \in S_n$, is a collection of three adjacent flat n-simplices

$$\{s_D \bmod \sigma, s \bmod \sigma, s_U \bmod \sigma\},$$

Where

$$s_D = ax_{\rho(1)} \left[x_{\rho(2)} \cdots x_{\rho(n-1)} x_{\rho(1)} \right]$$

And

$$s_U = a \left[x_{\rho(1)} \cdots x_{\rho(n-2)} x_{\rho(n)} \right].$$

For example, $[x_1 x_2] \bmod \sigma^2 \in TB_2$ specifies local trajectory $\{x_1[x_2 x_1] \bmod \sigma, [x_1 x_2] \bmod \sigma, [x_1 x_3] \bmod \sigma\}$ at $[x_1 x_2] \bmod \sigma \in B_2$ (Figure 1(c)). We would obtain a flow on B_n by patching these local trajectories together.

To define the "second derivative" along a trajectory, we would impose a kind of "smoothness condition" on local trajectories.

Definition 7: (Smoothness condition): Let Γ be a section of TB_n on $\{s_D \bmod \sigma, s \bmod \sigma, s_U \bmod \sigma\} \subset B_n$

Where

$$s = a\left[x_{\rho(1)} \cdots x_{\rho(n-1)}\right] \in S_n.$$

Suppose that

$$D\left[\Gamma(s \bmod \sigma)\right] = e/x_{\rho(n)}.$$

Then, we impose the following conditions on the local trajectory:

$$\begin{cases} D\left[\Gamma(s_D \bmod \sigma)\right] = e/x_{\rho(n)} \ \text{or} \ e/x_{\rho(1)}, \\ D\left[\Gamma(s_U \bmod \sigma)\right] = e/x_{\rho(n)} \ \text{or} \ e/x_{\rho(n-1)}. \end{cases}$$

Remark 8: For any two consecutive n-simplices $\{t_1, t_2\}$ on a trajectory, there exist a $a \in L_n$ and $\rho \in Sym_n$

s.t.

$$t_1 = a\left[x_{\rho(1)} \cdots x_{\rho(n-2)} x_{\rho(n-1)}\right]$$

And

$$t_2 = a\left[x_{\rho(1)} \cdots x_{\rho(n-2)} x_{\rho(n)}\right] \bmod \sigma.$$

Monomial $x_{\rho(1)} \dots x_{\rho(n-2)}$ is uniquely determined by $\{t_1, t_2\}$ and is included in both $D[\Gamma(t_1)]$ and $D[\Gamma(t_2)]$ for any section Γ of TB_n on $\{t_1, t_2\}$. That is, $x_{\rho(1)} \dots x_{\rho(n-2)}$ corresponds to the contact surface between two consecutive slant n-simplices.

As an example, let's consider the case of $n = 2$ shown in Figure 1(d), where the gradient at current triangle $a[x_1 x_2] \bmod \sigma$ is $x_1 x_2$. Then, the gradient at next triangle $ax_1[x_2x_1] \mathrm{mod}\, \sigma$ could assume either x_1x_2 or x_2x_3. Otherwise, we couldn't connect the two consecutive slant triangles over the trajectory "smoothly" as shown in the figure.

Tangent Cone and Section of TBn

Now we give the definition of the "peaks and valleys" of n-simplices (Figure 1(a)).

Definition 9: For $A \subset L_n$, tangent cone ConeA of L_n is defined as follows:

$$\mathrm{Cone}\, A = \left\{ px_1^{l_1} x_2^{l_2} \cdots x_n^{l_n} \,\middle|\, p \in A \text{ and } l_i \geq 0 \ (1 \leq i \leq n) \right\}.$$

Definition 10: For tangent cone $w = \mathrm{Cone}A$ $(A \subset L_n)$, boundary surfaces $d_s w$ is defined by

$$d_S w = \left\{ a\left[x_{\rho(1)} \cdots x_{\rho(n-1)} \right] \in S_n \,\middle|\, l_w(a_i) = 0 \ (0 \leq i < n) \right\},$$

Where

$$a_0 = a, a_1 = ax_{\rho(1)}, \cdots, a_{n-1} = ax_{\rho(1)} x_{\rho(2)} \cdots x_{\rho(n-1)} \in \mathbb{R}^n$$

and, for $z \in L_n$,

$$l_w(z) = \max_{p \in w} \left\{ \min \left\{ l_1, l_2, \cdots, l_n \right\} \middle| l_i \in \mathbb{Z} \ (1 \leq i \leq n) \ \text{s.t.} \ x_1^{l_1} x_2^{l_2} \cdots x_n^{l_n} = z/p \right\} \right\}.$$

Then, $d_s w$ specifies a unique slant n-simplex over each $t \in B_n$ and we obtain a section of S_n on B_n.

Definition 11: Γ_w is the section of S_n on B_n induced by tangent cone w, i.e., for $t \in B_n$, $\Gamma_w(t) = s \in d_s w$ s.t. $t = s \,\mathrm{mod}\, \sigma$.

Note that tangent cone w induces a section of TB_n on B_n by $DT_w : B_n \to \left\{ e/x_1, e/x_2, \cdots, e/x_n \right\}$.

Patching the local trajectories specified by $D\Gamma_w$ together, we would obtain a flow on B_n. As an example, let's consider the "peaks and valleys" shown in Figure 1(a), which is induced by

$$w = \text{Cone}\left\{1, x_1 x_2^{-1}, x_1^2 x_2 x_3^{-1}\right\}.$$

Let's start from triangle $[x_1 x_2] \bmod \sigma$ (grey) and move downward (Figure 2): $t[0] = [x_1 x_2] \bmod \sigma$ and $D\Gamma_w(t[0]) = x_1 x_2$. $D\Gamma_w(t[0])$ Specifies local trajectory $\{x_1[x_2 x_1] \bmod \sigma, [x_1 x_2] \bmod \sigma, [x_1 x_3] \bmod \sigma\}$ at $t[0]$.

Since we move downward, next triangle $t[1]$ is $x_1[x_2 x_1] \bmod \sigma$ and we obtain $D\Gamma_w\big(t[1]\big) = x_1 x_2$.

Then,

$D\Gamma_w(t[1])$ specifies local trajectory $\{x_1 x_2[x_1 x_2] \bmod \sigma, x_1[x_2 x_1] \bmod \sigma, [x_1 x_2] \bmod \sigma\}$ at $t[1]$ and next triangle $t[2]$ is $x_1 x_2[x_1 x_2] \bmod \sigma$. Continuing the process, we obtain a closed trajectory of length 10.

Finally, we consider variation of gradient, i.e., "second derivative", along a trajectory. Thanks for the smoothness condition, variation of gradient along a trajectory could be specified as a binary valued sequence.

Definition 12: Let $\{t[i]\} \subset B_n$ be a trajectory induced by $D\Gamma_w$ for tangent cone w. Then, "second derivative" $D^2\Gamma_w$ of Γ_w along $\{t[i]\}$ is defined as a $\{U, D\}$-valued function:

$$D^2\Gamma_w\big(t[i+1]\big) = \begin{cases} D^2\Gamma_w\big(t[i]\big) & \text{if } D\Gamma_w\big(t[i+1]\big) = D\Gamma_w\big(t[i]\big), \\ -D^2\Gamma_w\big(t[i]\big) & \text{else,} \end{cases}$$

Where $-D = U$ and $-U = D$.

Then, we could encode the conformation of a trajectory by the second derivative along the trajectory. As an example, let's consider the trajectory of Figure 2 again. First, set any initial value: $D^2\Gamma_w(t[0]) = D$

Then, since the first two triangles t[0] and t[1] have the same gradient, $D^2\Gamma_w(t[1]) = D$ The value of the second derivative is D until t[3], where it is changed to U because the gradient of t[2] is different from that of t[3]. Continuing the process, we obtain a binary sequence of length 10, DDDUDUUUDU, which describes the shape swept by the trajectory of triangles.

ENCODING OF PROTEIN BACKBONE STRUCTURE

In the case of n = 3, we obtain a flow of 3-simplices (tetrahedrons), which is used for protein structure analysis. In this section we propose a simple encoding method which translates the conformation of a protein backbone into a sequence of letters from a 16-letter alphabet (called D^2codes), using the second derivative along trajectories of tetrahedrons.

First, we consider all the fragments of five amino-acids occurred in a protein. Each fragment is approximated by a tetrahedron sequence of length five, where we permit translation and rotation during the process to absorb irregularity inherent in actual protein structures.

Next, we compute the second derivative along the tetrahedron sequences to obtain binary-valued sequences of length five. We assign the binary-valued sequences, which are denoted as a base-32 number, to the center amino-acid of the corresponding fragment. For example, DDDUD is denoted by "2", DUDDU is denoted by "9", DUDUD is denoted by "A", and so on.

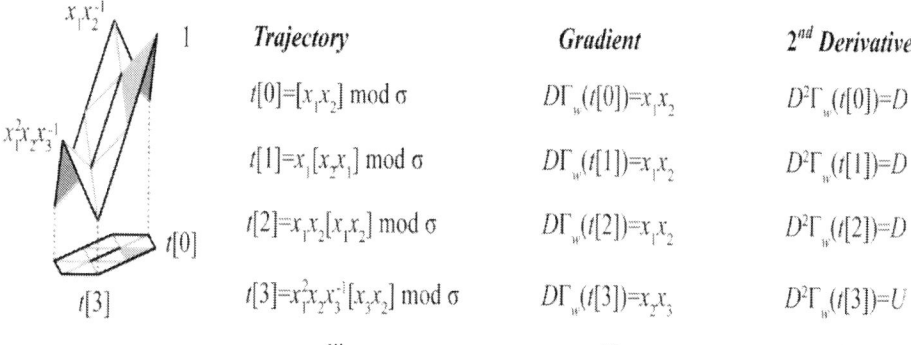

Figure 2: Closed trajectory of 2-simplices induced by $\text{Cone}\{1, x_1 x_2^{-1}, x_1^2 x_2 x_3^{-1}\}$.

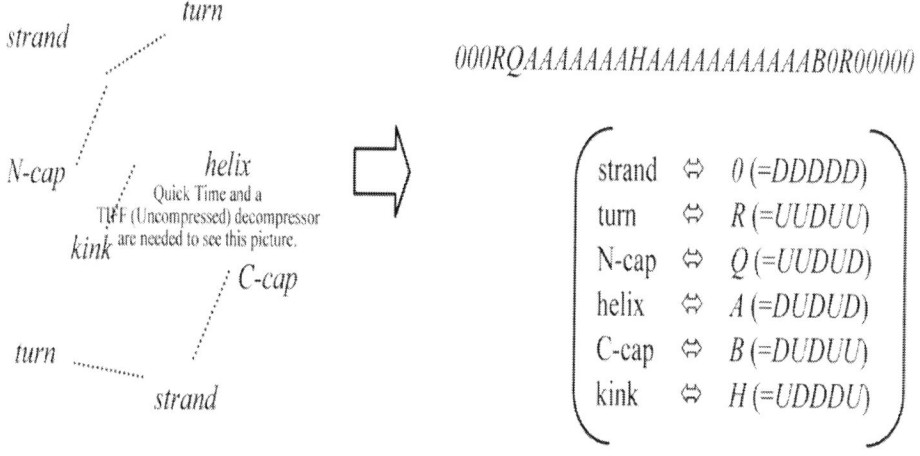

Figure 3: D^2-encoding of a protein (transferase 1RKL).

Then, we obtain a one-dimensional representation of protein backbone structure by arranging the base-32 numbers in the order the corresponding amino-acids appear in the protein. See [1] for detailed description of the algorithm.

Figure 3 shows an example of D^2-encoding of a protein. As you see, our method captures successfully not only recurring structural features of the protein (strand, turn, caps, helix), but also distortions (such as kink) as well.

DISCUSSION

In this paper, we first describe the discrete differential geometry of n-simplices. Then, we apply the mathematical framework to analysis of protein structures and propose a simple encoding method which translates the conformation of a protein backbone into a 16-valued sequence.

Unlike previous works, our version of discrete differential geometry studies connection between space-filling n-simplices. Considering cones of an integer lattice, we have introduced tangent bundle-like structure on n-simplices naturally. On notable consequence is the smoothness condition, i.e., restriction on variation of gradient along a trajectory. In particular, we could encode the shape of n-dimensional objects if we approximate them by sweeping the occupied area with a trajectory of n-simplices.

As for protein structure analysis, since we do not use clustering analysis to encode local structures, our approach not only provides a intuitively understandable description of protein structures, but also covers wide varieties of distortions. Our method performs comparably with more sophisticated but more time-consuming methods which are specifically designed for protein structure analysis. In SHREC'10 Protein Model Classification we achieved results comparable to more sophisticated methods, using the length of the longest common subsequence as the measure of structural similarity [12]. At homology level of CATH95 data set, our method performs best among all the individual classifiers considered in [13].

REFERENCES

1. Morikawa, N. (2007) Discrete Differential Geometry of Tetrahedrons and Encoding of Local Protein Structure. arXiv: math.CO/0710.4596.
2. Morikawa, N. (2011) A Novel Method for Identification of Local Conformational Changes in Proteins. arXiv: q-bio. BM/1110.6250.
3. Bobenko, A.I. and Suris, Yu.B. (2008) Discrete Differential Geometry. Integrable Structure. Graduate Studies in Mathematics, 98, 404 p. arXiv:math/0504358.

4. Meyer, M., Desbrun, M., Schroder, P. and Barr, A.H. (2003) Discrete Differential-Geometry Operators for Triangulated 2-Manifolds. In: Hege, H.-C. and Polthier, K., Eds., Visualization and Mathematics III, Springer-Verlag, Berlin, 35-58.http://dx.doi.org/10.1007/978-3-662-05105-4_2

5. Sillitoe, I., et al. (2013) New Functional Families (FunFams) in CATH to Improve the Mapping of Conserved Functional Sites to 3D Structures. Nucleic Acids Research, 41, D490-D498. http://dx.doi.org/10.1093/nar/gks1211

6. Murzin, A.G., Brenner, S.E., Hubbard, T. and Chothia, C. (1995) SCOP: A Structural Classification of Proteins Database for the Investigation of Sequences and Structures. Journal of Molecular Biology, 247, 536-540. http://dx.doi.org/10.1016/S0022-2836(05)80134-2

7. Rackovsky, S. and Scheraga, H.A. (1978) Differential Geometry and Polymer Conformation. 1. Comparison of Protein Conformations. Macromolecules, 11, 1168-1174.http://dx.doi.org/10.1021/ma60066a020

8. Louie, A.H. and Somorjai, R.L. (1982) Differential Geometry of Proteins: A Structural and Dynamical Representation of Patterns. Journal of Theoretical Biology, 98, 189-209.http://dx.doi.org/10.1016/0022-5193(82)90258-2

9. Montalvao, R.W., Smith, R.E., Lovell, S.C. and Blundell, T.L. (2005) CHORAL: A Differential Geometry Approach to the Prediction of the Cores of Protein Structures. Bioinformatics, 21, 3719-3725. http://dx.doi.org/10.1093/bioinformatics/bti595

10. Gorielyn, A., Hausrath, A. and Neukirch, S. (2008) The Differential Geometry of Proteins and Its Applications to Structure Determination. Biophysical Reviews and Letters, 3, 77-101. http://dx.doi.org/10.1142/S1793048008000629

11. Hu, S., Lundgren, M. and Niemi, A.J. (2011). The Discrete Frenet Frame, Inflection Point Solitons and Curve Visualization with Applications to Folded Proteins. arXiv:q-bio.BM/1102.5658.

12. Mavridis, L., et al. (2010) SHREC'10 Track: Protein Model Classification. Proceedings of Eurographics Workshop on 3D Object Retrieval, 117-124.

13. Boujenfa, K. and Limam, M. (2012) Consensus Decision for Protein Structure Classification. Journal of Intelligent Learning Systems and Applications, 4, 216-222.http://dx.doi.org/10.4236/jilsa.2012.43022

CITATION

Morikawa, N. (2014) Discrete Differential Geometry of n-Simplices and Protein Structure Analysis. Applied Mathematics, 5, 2458-2463. doi: 10.4236/am.2014.516237.

New Practical Algebraic Public-Key Cryptosystem and Some Related Algebraic and Computational Aspects

S. K. Rososhek

Faculty of Mathematics and Mechanics,
Tomsk State University, Tomsk, Russia

ABSTRACT

The most popular present-day public-key cryptosystems are RSA and ElGamal cryptosystems. Some practical algebraic generalization of the ElGamal cryptosystem is considered-basic modular matrix cryptosystem (BMMC) over the modular matrix ring $M_2(\mathbb{Z}_n)$. An example of computation for an artificially small number n is presented. Some possible attacks on the cryptosystem and mathematical problems, the solution of which are necessary for implementing these attacks, are studied. For a small number n, computational time for compromising some present-day public-key cryptosystems such as RSA, ElGamal, and Rabin, is compared with the corresponding time for the **BMMC**. Finally, some open mathematical and computational problems are formulated.

INTRODUCTION

Security of some present-day public-key cryptosystems is based on computational complexity of some numbertheoretical problems. Two of these problems are used most often: the integer factorization problem and the discrete logarithm problem. These problems ensure the security of the RSA and ElGamal cryptosystems, as well as of the corresponding digital signature schemes [1].

However, the true level of the computational complexity of these problems is unknown. That is to say, they are widely believed to be intractable, although no proof of this fact is known.

In [2], randomized polynomial-time algorithms for computing discrete logarithms and integer factoring were presented for the quantum computer.

Nevertheless, some alternatives should be proposed. One of possible approaches is to replace number-theoretical cryptosystems by such algebraic cryptosystems that would be resistant to an attack on a quantum computer.

Let us now consider some scheme of cryptosystems, namely, cryptosystems of group rings.

In the author's work [3, 4], a scheme of group ring cryptosystems was proposed. The idea to apply group rings in cryptography is based on the fact that if we fix the cardinality of a finite ring R, the cardinality of the group ring RG for a finite group G is an exponent of the cardinality of the group G. Then, a legal user can perform cryptographic transformations separately in the ring R and in the group G using polynomial algorithms and the illegal user has to solve computationally difficult problems in the group ring RG.

Let us consider the standardization problem in the group ring and two its aspects. The direct standardization problem is to construct a standard automorphism g of the group ring RG from an automorphism a of the group G and automorphism b of the ring R in the following way: if an element x of the group ring RG is represented as a formal linear combination of elements g_i of the group G with coefficients r_i from the ring R, then the image of the element x under the action of g is a formal linear combination of images of the elements g_i of the group G under the action of a with coefficients that are images of the coefficients r_i under the action of b.

The inverse standardization problem is formulated as follows. For a given automorphism g of a group ring RG, find an automorphism a of the group G and an automorphism b of the ring R such that g can be constructed from a and b by the way that was mentioned in the direct standardization problem or prove that such automorphisms a and b do not exist.

It is easy to see that, in the case of an efficient specification of the automorphism a in the group G and of the automorphism b in the ring R, one can efficiently compute the action of the automorphism g on any element of the group ring RG, i.e., efficiently specify the automorphism g of the ring RG.

As for the inverse standardization problem, there are some reasons to believe that this problem is computationally difficult. However, there is no proof for this statement.

In [5] some generalization of group ring cryptosystem is considered in the case of quasi group ring.

The question "For which finite commutative rings R and finite groups G all automorphisms of the group ring RG are standard automorphisms?" was partially answered in [6-8]. It should be noted that an inner automorphism of an integral group ring of a finite group is not a standard automorphism as a rule. This is why, together with the standard automorphisms of the group ring $\mathbb{Z}G$, where G is a finite group, we use inner automorphisms. In [9] the group ring $\mathbb{Z}S_3$, where S_3 is the permutation group for three symbols, is represented in a matrix form as block diagonal matrices of the fourth degree with two one-dimensional blocks and one two-dimensional block. In [9, 10] it is shown that the unit group of the group ring $\mathbb{Z}S_3$ is a semi-direct product of trivial units $(\pm S_3)$ and a free subgroup of rank 3. Since matrices of the fourth degree from this subgroup contain two identity one-dimensional blocks, we can restrict ourselves by a free group of matrices of the second degree with the free generators [9]:

$$A = \begin{pmatrix} 1 & 0 \\ 3 & 1 \end{pmatrix}, B = \begin{pmatrix} 1 & 3 \\ 0 & 1 \end{pmatrix}, C = \begin{pmatrix} -2 & 3 \\ -3 & 4 \end{pmatrix}$$

If we fall outside the limits of the matrix representation of $\mathbb{Z}S_3$, we consider arbitrary matrices of the second degree from the ring $M_2(\mathbb{Z})$ and its unit group $GL_2(\mathbb{Z})$, which contains free rank 3 subgroups $G(\alpha,\beta,\gamma)$ with the free generators

$$A(\alpha) = \begin{pmatrix} 1 & 0 \\ \alpha & 1 \end{pmatrix}, B(\beta) = \begin{pmatrix} 1 & \beta \\ 0 & 1 \end{pmatrix},$$

$$C(\gamma) = \left(\begin{array}{c|c} 1-\gamma & \gamma \\ \hline -\gamma & \gamma+1 \end{array} \right),$$

Where $\alpha,\beta,\gamma \in \mathbb{Z}$ and $|\alpha| \geq 3, |\beta| \geq 3, |\gamma| \geq 3$. [11]. For example, if $\alpha = \beta = \gamma = 3,$, we obtain a free rank 3 subgroup G=G (3, 3, 3) with the aforesaid free generators A, B, and C.

It should be also noted that all automorphisms of the group ring $\mathbb{Z}S_3$ are inner [12].

New practical algebraic generalization of the ElGamal cryptosystem will be given in the Section 2, some attacks on this cryptosystem—in the Section 4, new hard computational problems—in the Section 5, comparison of the security level of classical RSA, ElGamal and Rabin cryptosystems with security level of this cryptosystem for the same small number—in the Section 7, some related open mathematical and computational problems— in the Section 8. It should be noted, that some other theoretical algebraic generalizations of the ElGamal cryptosystem are given in [13, 14].

BASIC MODULAR MATRIX CRYPTOSYSTEM (BMMC)

Key Generation
User A does the following:

1. Picks large random positive integer n;

2. Picks the random words W(X) and W(U) in the alphabet $A^{\pm 1}, B^{\pm 1}, C^{\pm 1}$ in a free rank 3 group with free generators A, B, and C;

3. Computes the noncommuting matrices X_n, U_n by replacing the symbols A, B, and C in the words $W(X)$ and $W(U)$ by the corresponding matrices

$$A = \begin{pmatrix} 1 & 0 \\ 3 & 1 \end{pmatrix}, B = \begin{pmatrix} 1 & 3 \\ 0 & 1 \end{pmatrix}, C = \begin{pmatrix} -2 & 3 \\ -3 & 4 \end{pmatrix}$$

And performing matrix computations modulo n, i.e.,

$$X_n = X(\bmod n), U_n = U(\bmod n);$$

If X_n and U_n commute, then return to 2);

4. Let f(n) be the cardinality of the group $GL_2(\mathbb{Z}_n)$ over $\mathbb{Z}_n$-residue ring modulo n, then user A picks the random integers

$$-f(n) < k < f(n), -f(n) < s < f(n), 1 < \ell < f(n);$$

5. Public key of user A is

$$(n, P_1, P_2, P_3) = \left(n, X_n, U_n^{-s} X_n^k U_n^s, U_n^\ell\right)$$

and its private key is (U_n, s, k).

Remark 1: Orders of matrices

$$A = \begin{pmatrix} 1 & 0 \\ 3 & 1 \end{pmatrix}, B = \begin{pmatrix} 1 & 3 \\ 0 & 1 \end{pmatrix}$$

in the group $GL_2(\mathbb{Z}_n)$ are equal to n;

Remark 2: The cardinality of the group $GL_2(\mathbb{Z}_n)$ in the case $n=p^i$, p is a prime number, i is a positive integer, is equals to

$$f(p^i) = \left|GL_2\left(\mathbb{Z}_{p^i}\right)\right| = p^{4i-3}(p^2-1)(p-1) \quad [15].$$

As consequence in the case n=pq, p, q are primes, we have

$$f(pq) = \left| GL_2\left(\mathbb{Z}_{pq}\right) \right|$$
$$= p\left(p^2 - 1\right)\left(p - 1\right)q\left(q^2 - 1\right)\left(q - 1\right).$$

Encryption
User B does the following:

1. Writes the plaintext as a sequence of N numbers from $\mathbb{Z}_n$, where N is a multiple of 4, $\ell_1, \ell_2, ..., \ell_N$, adding, if necessary, numbers from the first quadruple by a cyclic permutation at the end of the sequence;

2. Writes each quadruple of numbers of the obtained sequence similarly as matrix:

$$m^{(1)} = \begin{pmatrix} \ell_1 & \ell_2 \\ \ell_3 & \ell_4 \end{pmatrix} \in M_2\left(\mathbb{Z}_n\right);$$

3. Picks session keys-random integers $r_i, t_i, -f(n) < r_i < f(n), -f(n) < t_i < f(n)$ for each of $N/4$ obtained matrices $m^{(i)}$;

4. Computes the ciphertext block for each matrix $m^{(i)}$:

$$\left(C_1^{(i)}, C_2^{(i)}\right) = \left(P_3^{-r_i} P_1^{t_i} P_3^{r_i}, m^{(i)} P_3^{-r_i} P_2^{-t_i} P_3^{r_i}\right)$$
$$i = 1, 2, \cdots, N/4.$$

Decryption
Using the private key, user A computes for each ciphertext block $\left(C_1^{(i)}, C_2^{(i)}\right)$:

$$C_2^{(i)} U_n^{-s} \left(C_1^{(i)}\right)^k U_n^s = m^{(i)}.$$

After obtaining the sequence of matrices

$$m^{(1)}, m^{(2)}, \cdots, m^{(N/4)},$$

the sequence of numbers $\ell_1, \ell_2, ..., \ell_N$ and hence the plaintext can be reconstructed uniquely.

Theorem: Decryption in the BMMC is correct.

Proof: It is sufficiently to consider a case of one block of the ciphertext:

$$C_2 U_n^{-s} C_1^k U_n^s$$

$$= \left(m P_3^{-r} P_2^{-t} P_3^r \right) U_n^{-s} \left(P_3^{-r} P_1^t P_3^r \right)^k U_n^s$$

$$= m P_3^{-r} P_2^{-t} P_3^r U_n^{-s} P_3^{-r} P_1^{kt} P_3^r U_n^s$$

$$= m U_n^{-r\ell} U_n^{-s} X_n^{-kt} U_n^s U_n^{r\ell} U_n^{-s} U_n^{-r\ell} X_n^{kt} U_n^{r\ell} U_n^s$$

$$= m \left(U_n^{-r\ell-s} X_n^{-kt} U_n^0 X_n^{kt} U_n^{r\ell+s} \right)$$

$$= m \left(U_n^{-r\ell-s} U_n^{r\ell+s} \right) = m.$$

It should be noted that algorithms of the BMMC are implemented using the algorithm of matrix modular exponentiation similar to the usual modular exponentiation algorithm in which multiplication of integers is replaced by multiplication of matrices with reduction of their elements modulo n. In addition parallel computations may be used in matrix multiplications to increase the computational efficiency of the cryptosystem.

Let n be a large 256 bit integer, then the cardinality bit length of the group $GL_2(\mathbb{Z}_n)$ would be near 800 bits or more. For comparing in the case of the ElGamal cryptosystem the bit lengths of p and the cardinality of corresponding multiplicative group of residue field $\mathbb{Z}_p$ are equal. But one reduction modulo 1024 bit number in the ElGamal cryptosystem costs as some reductions modulo 256 bit number in the BMMC. Therefore, under corresponding choice of parameters the BMMC may be faster than the ElGamal cryptosystem with the same security level, because the gybrid problem and the transformation problem are harder than the discrete logarithm problem in the groups of the same cardinality.

EXAMPLE

Key Generation

User A does the following:

1. Picks two prime numbers p=17 and q=19 and computes $n = 17 \times 19 = 323$;

2. Picks the words in the free group:

$$W(U) = C^{-4}A^3B, W(X) = B^{-1}C;$$

3. Computes matrices modulo n:

$$U_n = U(\mathrm{mod}\, n)$$

$$= \left[\left(\begin{array}{c|c} 321 & 3 \\ \hline 320 & 4 \end{array} \right)^{-4} \left(\begin{array}{c|c} 1 & 0 \\ \hline 3 & 1 \end{array} \right)^{3} \left(\begin{array}{c|c} 1 & 3 \\ \hline 0 & 1 \end{array} \right) \right] (\mathrm{mod}\, 323)$$

$$= \left(\begin{array}{c|c} 228 & 26 \\ \hline 236 & 51 \end{array} \right),$$

$$X_n = X(\mathrm{mod}\, n)$$

$$= \left[\left(\begin{array}{c|c} 1 & 3 \\ \hline 0 & 1 \end{array} \right)^{-1} \left(\begin{array}{c|c} 321 & 3 \\ \hline 320 & 4 \end{array} \right) \right] (\mathrm{mod}\, 323)$$

$$= \left(\begin{array}{c|c} 7 & 314 \\ \hline 320 & 4 \end{array} \right);$$

matrices U_n and X_n do not commute and, therefore, the user passes to the next step;

4. Picks the integers $k=s=1, \ell = 2$

5. The public key is

$$(n, P_1, P_2, P_3)$$

$$= \left(n = 323, \left(\begin{array}{c|c} 7 & 314 \\ \hline 320 & 4 \end{array} \right), \left(\begin{array}{c|c} 227 & 39 \\ \hline 101 & 107 \end{array} \right), \left(\begin{array}{c|c} 303 & 148 \\ \hline 275 & 16 \end{array} \right) \right);$$

the private key is $\left(\left(\begin{smallmatrix} 228 \\ 236 \end{smallmatrix} + \begin{smallmatrix} 26 \\ 51 \end{smallmatrix} \right), k = s = 1 \right).$

Encryption
User B does the following:

1. Writes the plaintext as a sequence of numbers from $\mathbb{Z}_n$. The length of this sequence is multiple of 4. If necessary, some numbers are added. For example, let the plaintext be

$$4\|5\|7\|8\|17\|15\|10\|;$$

Here, a number should be added to the last block by shifting the first number cyclically, the user obtains two quadruples of numbers from $\mathbb{Z}_n$:

$$4\|5\|7\|8,17\|15\|10\|4;$$

2. Writes the plaintext as two matrices from $M_2(\mathbb{Z}_n)$:

$$m^{(1)} = \left(\begin{array}{c|c} 4 & 5 \\ \hline 7 & 8 \end{array} \right), m^{(2)} = \left(\begin{array}{c|c} 17 & 15 \\ \hline 10 & 4 \end{array} \right);$$

3. Encrypts each block (matrix) separately choosing different session keys. For example, the first block is encrypted as follows;

4. Picks the session key for the first block $r_1 = 2$, $t_1 = 1$;

5. Computes the ciphertext of the first block modulo $n=323$:

$$C_1^{(1)} = P_3^{-2} P_1 P_3^2 \bmod n = \left(\begin{array}{c|c} 303 & 148 \\ \hline 275 & 16 \end{array} \right)^{-2} \left(\begin{array}{c|c} 7 & 314 \\ \hline 320 & 4 \end{array} \right) \left(\begin{array}{c|c} 303 & 148 \\ \hline 275 & 16 \end{array} \right)^2 (\bmod n) = \left(\begin{array}{c|c} 220 & 245 \\ \hline 105 & 114 \end{array} \right)$$

$$C_2^{(1)} = m^{(1)} \cdot P_3^{-2} P_2^{-1} P_3^2 = \left(\begin{array}{c|c} 4 & 5 \\ \hline 7 & 8 \end{array} \right) \left(\begin{array}{c|c} 303 & 148 \\ \hline 275 & 16 \end{array} \right)^{-2} \left(\begin{array}{c|c} 107 & 284 \\ \hline 202 & 227 \end{array} \right) \left(\begin{array}{c|c} 303 & 148 \\ \hline 275 & 16 \end{array} \right)^2 (\bmod n) = \left(\begin{array}{c|c} 217 & 206 \\ \hline 205 & 13 \end{array} \right).$$

The ciphertext of the second block $M^{(2)}$ is computed similarly with the choice of another session key r_2, t_2.

Decryption

User A, having obtained the ciphertext from user B, does the following: using its private key, for each ith block, computes

$$C_2^{(i)} U_n^{-1} C_1^{(i)} U_n;$$

In particular, for the first block, he obtains

$$C_2^{(1)} U_n^{-1} C_1^{(1)} U_n = \left[\left(\frac{217 \mid 206}{205 \mid 13} \right) \left(\frac{51 \mid 297}{87 \mid 228} \right) \left(\frac{220 \mid 245}{105 \mid 114} \right) \left(\frac{228 \mid 26}{236 \mid 51} \right) \right] \bmod n$$

$$= \left(\frac{217 \mid 206}{205 \mid 13} \right) \left(\frac{248 \mid 97}{90 \mid 86} \right) \bmod n = \left(\frac{4 \mid 5}{7 \mid 8} \right) = m^{(1)}.$$

SOME ATTACKS ON BMMC

Find the Private Key (Un, s, k) by the Public Key (n, P1, P2, P3)

1. Let the cardinality of the group $GL_2(\mathbb{Z}_n)$ be

$$|GL_2(\mathbb{Z}_n)| = f(n);$$

Since $P_3 = U_n^\ell$, the cryptanalyst can try to solve the equation with two unknowns Y and x:

$$Y^x = P_3,$$

Where $Y, P_3 \in GL_2(\mathbb{Z}_n), -f(n) < x < f(n)$.

2. Since

$$P_2 = U_n^{-s} P_1^k U_n^s,$$

the cryptanalyst can try to solve the equation with two unknowns Z and x:

$$Z P_2 Z^{-1} = P_1^x,$$

Where $Z, \in GL_2(\mathbb{Z}_n), -f(n) < x < f(n)$, what leads to the private key by applying 1) to each solution Z_0 (which we call the transforming matrix).

Find the Private Key (Un, s, k) by the Ciphertext (C1, C2)

Since the private key is applied in the ciphertext (C_1, C_2) not directly but only via the public key, the knowing of only the ciphertext does not yield additional possibilities to the attacks from 4.1 for the attack on the private key.

Find the Session Key (r, t) by the Ciphertext (C1, C2)

1. Since $C_1 = P_3^{-r} P_1^t P_3^r$, the cryptanalyst can try to solve the equation with two unknowns Z and y:

$$ZC_1Z^{-1} = P_1^y,$$

Where $Z, \in GL_2(\mathbb{Z}_n), -f(n) < y < f(n),$

2. For any solution (Z_0, y_0) of the equation from 1), the cryptanalyst can try to solve the equation with two unknowns Y and x:

$$Y^x = Z_0,$$

Where $Y, Z_0, \in GL_2(\mathbb{Z}_n), -f(n) < x < f(n),$

Find the Corresponding Plaintext m or the Session Key (r, t) by a Chosen Ciphertext (C1, C2)

Cryptanalyst chooses the random $\tilde{m} \in GL_2(\mathbb{Z}_n)$ and computes $\tilde{m}C_2$, then send it to user A for decryption. User A computes:

$$\left(\tilde{m}C_2\right)U_n^{-s}C_1^{-k}U_n^s = \tilde{m}\left(C_2U_n^{-s}C_1^{-k}U_n^s\right) = \tilde{m}m.$$

and send the result to cryptanalyst, which computes the plaintext:

$$\tilde{m}^{-1}\left(\tilde{m}m\right) = m.$$

Hence for protecting cryptosystem the modification of encryption algorithm is:

$$C_2 P_3^{-r} P_2^{-t} P_3^r m P_3^{-r} P_2^t P_3^r,$$

the modification of decryption algorithm is:

$$U_n^{-s}C_1^kU_n^sC_2U_n^{-s}C_1^kU_n^s = m.$$

COMPUTATIONAL PROBLEMS IN ENSURING BMMC SECURITY

From the consideration of attacks 4.1-4.4 one can formulate some problems, the solution of which is necessary to implement the corresponding attacks.

The Transformation Problem

Let a matrix P_2 be conjugated with an unknown integral power of a matrix P_1 for two given matrices $P_1, P_2, \in GL_2(\mathbb{Z}_n)$. Find all solutions of the equation with two unknowns Z and y:

$$ZP_2Z^{-1} = P_1^y,$$

Where $Z \in GL_2(\mathbb{Z}_n), -f(n) < y < f(n)$.

Let us consider a particular case of Problem 5.1.

1. The conjugation problem:

   For two given conjugated matrices P_2 and $P_1^{y_0}$ from the group GL_2 $(\mathbb{Z}_n)$, find a transforming matrix $T \in GL_2(\mathbb{Z}_n)$, i.e., matrix T such that

   $$T^{-1}P_2T = P_1^{y_0}$$

The Hybrid Problem

Find all solutions of the equation with two unknowns Y and x

$$Y^x = Z_0,$$

Where $Y, Z_0 \in GL_2(\mathbb{Z}_n), -f(n) < y < f(n)$ in the group $GL_2(\mathbb{Z}_n)$.

Let us also consider two particular cases of Problem 5.2.

1. The discrete logarithm problem in a cyclic subgroup of the group $GL_2(\mathbb{Z}_n)$.

   Let $H = \langle Y_0 \rangle$ be a fixed cyclic subgroup of order j of the group GL_2 $(\mathbb{Z}_n)$ with the generator Y_0, $M \in H$ be an arbitrary element. Find the unique solution $x = x_0$ of the equation

   $$Y^2 = M,$$

   Where x is an integer such that $0 \le x < j$.

1. The problem of extracting a root of the ith power in the group GL_2 $(\mathbb{Z}_n)$ (the matrix RSA problem).

   Let $M \in GL_2(\mathbb{Z}_n)$ be an arbitrary element, i_0 be a fixed integer satisfying the condition $0 \le i_0 < f(n)$ and $GCD(i_0, f(n)) = 1$.

   Find all solutions of the equation with a single unknown Y:

   $$Y^{i_0} = M, Y \in GL_2(\mathbb{Z}_n).$$

   According to the Problem 2), in turn, one can also discern the following problem.

   The problem of square-root extraction in $GL_2(\mathbb{Z}_n)$.

   Find all solutions of the equation with a single unknown Y: $Y^2 = M$

   Where $Y, M \in GL_2(\mathbb{Z}_n)$.

COMPUTATIONAL COMPLEXITY OF PROBLEMS 5.1, 5.2

If the order $O(P_1) = O(P_2) = j$ is a large number, then, the fact that the generators in a cyclic group are indistinguishable and random choice of k in the key generation show, on the one hand, that the identification of matrices P_1^y in Problem 5.1 is a hard problem and, on the other hand, the impossibility to implement the exhausting search in practice for a large number j.

Considering Problem 5.1 1), it should be noted that this problem is solvable in the free subgroup $G=G$ (3, 3, 3) of the group $GL_2(\mathbb{Z})$ (see [16]). The possibility to extend this algorithm for a subgroup of the group $GL_2(\mathbb{Z}_n)$ depends on the solution of the following problem: for a given matrix $X_n \in G_n \subset GL_2(\mathbb{Z}_n)$, find the word $W(X)$ and matrix $X \in G$ whose reduction modulo n yields the matrix X_n.

Nevertheless, even in the case of a solved problem of extension, the problem about the existence of an efficient algorithm for solving Problem 5.1 1) remains open.

Let us now consider Problem 5.2. As it is a problem with two unknowns, this problem is more complicated in the general case than its particular cases, the discrete logarithm problem and the problem of extracting a matrix root modulo n. It is worth to note that the square-root extracting problem is computationally difficult for large number N=pq, p and q are primes.

Let us now turn to the discussion of the cardinality of the set of secret keys for **BMMC**. Note that, for classical cryptosystems, the uniqueness of the secret key can be reached by fitting of parameters. For BMMC, the situation is other. Indeed, if a matrix T_0 transforms the matrix P_1^i into the matrix P_2, i.e.,

$$T_0^{-1} P_1^i T_0 = P_2,$$

then the matrix $Z_0 T_0$ also transforms P_1^i into P_2 for any matrix $Z_0 \in C(P_1)$, where $C(P_1)$ is a centralizer of P_1 in $GL_2(\mathbb{Z}_n)$, because

$$\left(T_0^{-1} Z_0^{-1}\right) P_1^i \left(Z_0 T_0\right) = T_0^{-1} \left(Z_0^{-1} Z_0\right) P_1^i T_0 = T_0^{-1} P_0^i T_0 = P_2.$$

Thus, if the secret key (U_n, s, k) is considered as initial, the cryptanalyst can compromise the BMMC by any real key of the form $\left(Z_0 U_n^s, k\right)$, where $Z_0 \in C(P_1)$. Then, for the cardinality of the set of real keys W_0, we have $W_0 \geq |C(P_1)|$ and when generating a key it is necessary to choose matrix P_1 so that $W_1 = \dfrac{W_0}{|GL_2(\mathbb{Z}_n)|}$ was negligibly small, e.g., $W_1 < 2^{-80}$.

This protects from random guessing of the private key.

COMPARISON OF COMPUTATIONAL SECURITY OF CLASSICAL RSA, ELGAMAL, AND RABIN CRYPTOSYSTEMS WITH BMMC

For demonstrativeness, we compare the cryptosystems for a very small number n=35.

RSA Cryptosystem
Let the public key be n=35, e=19).

In this case, the cryptanalyst instantaneously compromises RSA by factorization $n = 35 = 5 \times 7$, from which finds $\varphi(n) = 4 \times 6 = 24$, and computation of the secret key $d = e^{-1}(\operatorname{mod}\varphi(n))$ either by the extended Euclid's algorithm or by exhaustive search. Then the cryptanalyst finds the secret key:

$$d = 19^{-1}(\operatorname{mod} 24) = 19.$$

Modified ElGamal Cryptosystem

In the unit group $\mathbb{Z}_{35}^*$ of the ring $\mathbb{Z}_{35}$ one has to choose an element of the maximal order. For this purpose, n=35 is factorized as 35=5×7, and the generators are chosen in the groups $\mathbb{Z}_5^*$ and $\mathbb{Z}_7^*$, e.g.,

$$\mathbb{Z}_5^* = \langle 2 \rangle \quad \text{and} \quad \mathbb{Z}_7^* = \langle 3 \rangle.$$

Then the element of maximal order in $\mathbb{Z}_{35}^*$ is obtained from the solution of the following simultaneous congruences either by inspection or by the Chinese reminder theorem:

$$\begin{cases} x \equiv 2 \pmod 5, \\ x \equiv 3 \pmod 7. \end{cases}$$

It follows that x=17 and its order is 0(17) =12.

Let one of cyclic subgroups of order 12, for example, $G = \langle 17 \rangle$ be chosen in the group $\mathbb{Z}_{35}^*$. In the group G, another generator may be chosen, e.g., $G = \langle 3 \rangle$.

Let the modified ElGamal cryptosystem be considered in a cyclic group G of order 12 with a generator $\alpha = 3$ and let the public key be

$$\left(n = 35, \alpha = 3, \beta = \alpha^a \bmod 35 = 33 \right).$$

In this case, the cryptanalyst instantaneously compromises the modified ElGamal cryptosystem using exhaustive search in the cyclic group of order 12 finding the secret key a = 5 since $3^5 \pmod{35} = 33$.

Remark: In the case of choice n as n=p, where p is a prime number, we compare BMMC with classical ElGamal cryptosystem.

Rabin Cryptosystem
Let the public key be n= (35), then the cryptanalyst instantaneously compromises the Rabin cryptosystem in this case by factorizing the number by prime multipliers n=5×7.

One can see that, in all three cases, the cryptanalyst instantaneously compromises these classical cryptosystems for n=35. Let us now the case of the BMMC cryptosystem for n=35.

BMMC
Let the public key be

$$\left(n = 35, P_1 = \left(\begin{array}{c|c} 23 & 33 \\ \hline 12 & 34 \end{array} \right), P_2 = \left(\begin{array}{c|c} 31 & 0 \\ \hline 31 & 26 \end{array} \right), P_3 = \left(\begin{array}{c|c} 31 & 5 \\ \hline 15 & 16 \end{array} \right) \right),$$

$$(C_1, C_2) = \left(\left(\begin{array}{c|c} 14 & 12 \\ \hline 18 & 8 \end{array} \right), \left(\begin{array}{c|c} 5 & 18 \\ \hline 30 & 18 \end{array} \right) \right)$$

Be the cipher text of a certain matrix m.

Compromising BMMC in this case needs essentially more efforts than for the classical cryptosystems and exhausting search in the space of the search containing 775,760 matrices gives

$$m = \begin{pmatrix} 7 & 8 \\ \hline 2 & 3 \end{pmatrix},$$

Secret key—

$$U_n = \begin{pmatrix} 33 & 32 \\ \hline 26 & 21 \end{pmatrix},$$

k=17, s=5,l=3; session key—r=1, t=-1,.

SOME OPEN MATHEMATICAL AND COMPUTATIONAL PROBLEMS

1. For which finite groups G and rings R the unit group of group ring RG is a semi-direct product of trivial units and a free subgroup of a finite rank?
2. For which groups G and rings R every automorphism of the group ring RG has a standard form?
3. For which subgroups of the group $GL_2(\mathbb{Z}_n)$ it takes place the property of small centralizers i.e. every element has a cyclic centralizer?
   Remark: It is well-known [16] that in the free group of finite rank centralizer of any element is a cyclic subgroup.
4. Is there a polynomial-time algorithm for constructing cyclic centralizer of any element in a free group of finite rank?
5. Is there a polynomial-time algorithm for solving the membership problem for cyclic subgroup of the a) free group of finite rank, b) subgroup G=G (3, 3, 3) by modulo n in a group $GL_2(\mathbb{Z}_n)$?
6. Is there a polynomial-time algorithm for solving the modular factorization problem, i.e. to represent every matrix from the subgroup

$G=G$ (3, 3, 3) by modulo n in a group $GL_2(\mathbb{Z}_n)$ as a word in an alphabet of $A^{\pm 1}, B^{\pm 1}, C^{\pm 1}$ by modulo n?

7. How to compute the number f (n) for arbitrary positive integer's n? More exactly, is there a polynomialtime algorithm for computing f (n)?

8. Is there a polynomial-time algorithm for computing maximal order elements in a subgroup $G=G$ (3, 3, 3) by modulo n in the group GL_2 $(\mathbb{Z}_n)$? What is a cardinality of this subgroup G_n?

CONCLUSIONS

The practicality of the BMMC is provided by the absence of the necessity in the computer algebra systems used for computer realization of cryptosystem algorithms and efficient matrix computations by modulo number of essentially less bit length than that are usually used in classical cryptosystems under the same security level.

REFERENCES

1. Menezes, P. van Ooshot and S. Vanstone, "Handbook of Applied Cryptography," CRC Press, Waterloo, 1996. doi:10.1201/978143982191

2. P. W. Shor, "Algorithms for Quantum Computation: Discrete Logarithm and Factoring," Proceedings of the IEEE 35th Communications Annual Symposium on Foundations of Computer Science, Santa Fe, 20-22 November 1994, pp. 124-134.

3. S. K. Rososhek, "Cryptosystems in Automorphism Groups of Group Rings of Abelian Groups," Fundamentalnaya I prikladnaya matematica, Vol. 13, No. 8, 2007, pp. 157- 164 (in Russian).

4. S. K. Rososhek, "Cryptosystems in Automorphism Groups of Group Rings of Abelian Groups," Journal of Mathematical Sciences, Vol. 154, No. 3, 2008, pp. 386-391.doi:10.1007/s10958-008-9168-2

5. N. Gribov, P. A. Zolotykh and A. V. Mikhalev, "A Construction of Algebraic Cryptosystem over the Quasigroup Ring," Mathematical Aspects of Cryptography, Vol. 1, No. 4, 2010, pp. 23-32 (in Russian).

6. K. N. Ponomarev, "Automorphically Rigid Group Algebras I. Semisimple Algebras," Algebra and Logic, Vol. 48, No. 5, 2009, pp. 654-674. doi:10.1007/s10469-009-9064-y

7. K. N. Ponomarev, "Automorphically Rigid Group Algebras II. Modular Algebras," Algebra and Logic, Vol. 49, No. 2, 2010, pp. 216-237.
8. K. N. Ponomarev, "Rigid Group Rings," In: A. G. Pinus and K. N. Ponomarev, Eds., Algebra and Model Theory, 6, Novosobirsk Technical University Press, Novosibirsk, 2007, pp. 73-83 (in Russian). doi:10.1007/s10469-010-9086-5
9. Popova and E. Poroshenko, "Units Group of Integral Group Rings of Finite Groups," In: A. G. Pinus and K. N. Ponomarev, Eds., Algebra and Model Theory, 4, Novosibirsk Technical University Press, Novosibirsk, 2003, pp. 99-106 (in Russian).
10. Dooms and E. Jespers, "Normal Complements of the Trivial Units in the Unit Group of Some Integral Group Rings," Communications in Algebra, Vol. 31, No. 1, 2003, pp. 475-482. doi:10.1081/AGB-120016770
11. Y. I. Merzlyakov, "Matrix Representations of Free Groups," Doklady Akademii Nauk, Vol. 238, No. 3, 1978, pp. 527-533 (in Russian).
12. Popova, "Group of Automorphisms of the Ring $\mathbb{Z}S_3$," In: A. G. Pinus and K. N. Ponomarev, Eds., Algebra and Model Theory, 6, Novosibirsk Technical University Press, Novosibirsk, 2007, pp. 84-90 (in Russian).
13. Mahalanobis, "A Simple Generalization of the ElGamal Cryptosystem to Non-Abelian Groups," Communications in Algebra, Vol. 36, No. 10, 2008, pp. 3878-3889.doi:10.1080/00927870802160883
14. S.-H. Paeng, K.-C. Ha, J. N. Kim, S. Chee and C. Park, "New Public Key Cryptosystem Using Finite Non-Abelian Groups," Proceedings of the Crypto 2001, Lecture Notes in Computer Sciences, Santa Barbara, 19-23 August 2001, pp. 470-485.
15. M. I. Kargapolov and Y. I. Merzlyakov, "Foundations of Group Theory," Nauka, Moscow, 1977 (in Russian).
16. R. C. Lyndon and P. E. Schupp, "Combinatorial Group Theory," Springer-Verlag, Berlin, Heidelberg, New York, 1977.

CITATION

S. Rososhek, "New Practical Algebraic Public-Key Cryptosystem and Some Related Algebraic and Computational Aspects," Applied Mathematics, Vol. 4 No. 7, 2013, pp. 1043-1049. doi: 10.4236/am.2013.47142.

The Decomposition Theorem and the Intersection Cohomology of Quotients in Algebraic Geometry

Jonathan Woolf[1]
Christ's College, Cambridge, CB2 3BU, UK

ABSTRACT

Suppose a connected reductive complex algebraic group G acts linearly on an irreducible complex projective variety X. We prove that if

$$1 \to N \to G \to H \to 1$$

is a short exact sequence of connected reductive groups and X^{ss} the open set of semistable points for the action of N on X then $IH_H^*(X^{ss} \| N)$ is (non-canonically) a direct summand of $IH_G^*(X^{ss})$. The inclusion is provided by the decomposition theorem and certain resolutions of the action allow us to define projections.

INTRODUCTION

Suppose a connected complex reductive group G acts linearly on an ample line bundle over a complex projective variety X. There is a geometric invariant theory quotient $X^{ss} \| G$ of the set X^{ss} of semistable points of X for this linearisation. We assume that the open set X^s of stable points for the linearised action is non-empty in order that $\dim X^{ss} \| G = \dim X - \dim G$. When the variety X is smooth, and every semistable orbit is furthermore stable, the relation between the equivariant cohomology of X and the cohomology of $X^{ss} \| G$ has been intensively studied, see e.g. [12], [19], [8], [5] and [18] and many others. Aside from the intrinsic interest of relating an equivariant invariant to one defined on a quotient, the subject has obvious applications in the topological study of various moduli spaces. A key point in this theory is the observation that $H_G^*(X^{ss}) \cong H^*(X^{ss} \| G)$ under these assumptions. Several papers, for instance [14], [7] and [15] and [10], have studied what happens when we relax the assumption that every semistable orbit be stable. Since the quotient $X^{ss} \| G$ will in general then be singular they all consider its intersection cohomology groups rather than its ordinary cohomology. Principally this is because the former retain for singular projective varieties the structures, collectively known as the Kähler package, which hold for the cohomology of a smooth projective variety. So, although they are less tractable in terms of functoriality, in many ways they provide a richer invariant for the study of singular varieties. Another common theme in these papers is the construction of some resolution of the G action on X and the use of this to define a map $H_G^*(X^{ss}) \rightarrow IH^*(X^{ss} \| G)$ which is surjective. In this paper we analyse this approach in general terms and provide a framework in which to place their results.

Let us allow that the complex projective variety X may be singular and consider the quotient morphism $X^{ss} \rightarrow X^{ss} \| G$. We would like to relate the intersection cohomology $IH^*(X^{ss} \| G)$ of the quotient to the equivariant intersection cohomology $IH_G^*(X^{ss})$ of the semistable points. Throughout this paper we will take coefficients in the rationals $\mathbb{Q}$ and thus avoid issues of torsion. In slightly different terms we note that $IH_G^*(X^{ss})$ is, almost by definition, the intersection cohomology of the quotient stack $[X^{ss}/G]$ so that we are comparing the quotient stack with the geometric invariant theory quotient. When every semistable orbit is stable [2, Theorem 9.1] tells us that

$$IH_G^*(X^{ss}) \cong IH^*(X^{ss} \| G).$$

This of course corresponds to the fact that the quotient stack $[X^{ss}/G]$ is represented by the orbifold $X^{ss}/G \cong X^{ss} \| G$. If we remove this condition the picture is more complicated. Intuitively the equivariant intersection cohomology groups contain more information. This notion is expressed quite simply in our main result which says that $IH^*(X^{ss}||G)$ is always a direct summand of $IH^*_G(X^{ss})$. In fact we prove something slightly more general. Suppose N is a connected normal reductive subgroup of G, and H the corresponding quotient G/N. Let X^{ss} now stand for the set of N-semistable points of X. Then we show that $IH^*_H(X^{ss}||N)$ is a direct summand of $IH^*_G(X^{ss})$.

It is no great surprise that the key ingredient of our proof is the decomposition theorem of [1, Section 6], or rather its equivariant extension which was proved in [2, Section 5]. To put this result in context we review, in Section 1, the construction of the constructible equivariant derived category from [2]. Then in Section 2 we show how to define a map

$$IH^*_H(X^{ss}||N) \rightarrow IH^*_G(X^{ss}). \tag{1}$$

In the final section we consider certain resolutions of X^{ss}, which we term N-stable, and show how these can be used to construct projections

$$IH^*_G(X^{ss}) \rightarrow IH^*_H(X^{ss}||N)$$

and thence show that (1) is an inclusion.

THE EQUIVARIANT DERIVED CATEGORY

The results of this paper are couched in the language of the equivariant derived category introduced by Bernstein and Lunts in [2]. We give a very brief review of the structures which we use, but refer the reader to [2] for the details. Background material on the constructible derived category, t-structures, perverse sheaves and intersection cohomology can be found in [4] and [1].

Suppose that X is an irreducible complex algebraic variety. A 'sheaf on X' will be a module over the constant sheaf with coefficients in $\mathbb{Q}$, and will furthermore be constructible i.e. its cohomology sheaves will be locally constant on the strata of some stratification of X by smooth subvarieties. Such sheaves form an Abelian category Sh(X). We write D(X) for the bounded below derived category of constructible sheaves on X. The middle perversity intersection cohomology complex $\mathscr{IC}^\bullet$ (X) is defined up to quasi-isomorphism as an object of D(X) obeying axioms set out in [4]. We follow their definition except that we shift by the complex dimension d_X of X and so follow the Beilinson–Bernstein–Deligne–Gabber indexing used in[1] and [2] (see [4, 2.3] for a comparison of the various indexing systems). In addition to the standard t-structure on D(X), whose heart is Sh(X), there is a t-structure associated to the middle perversity whose heart is the Abelian category Perv(X) of perverse sheaves—see [2, Section 5] and [1, Section 2]. $\mathscr{IC}^\bullet$ (X) is a simple object in this full subcategory (see [1, 4.3]).

A map $\varphi : X - X'$ gives rise to functors $\varphi^*, \varphi^! : D(X') \to D(X)$ and $\varphi_*, \varphi_! : D(X) \to D(X')$. Here φ^* is the left adjoint of φ_* and $\varphi_!$ the left adjoint of $\varphi^!$. There is also a natural tensor product $\otimes$ on D(X).

Remark 1.1

Note that following [2] we write φ_* and not the more usual $R\varphi_*$ for the right derived functor because we will always work with derived categories and so there is no possibility of confusion with the pushforward of sheaves.

The intersection cohomology groups of X are defined by

$$IH^*(X) := H^{*+d_X}(\pi_* \mathscr{IC}^\bullet(X))$$

where $\pi : X \to \mathrm{pt}$ is the map to a point, and d_X the complex dimension of X.

Now suppose that a reductive algebraic group G acts algebraically on X. If X is a principal G-space then the derived category D(X/G) of the quotient is a good definition of 'equivariant derived category'. More generally let $\mathrm{Res}_G(X)$ be the category of G-resolutions of X i.e. the category whose objects are equivariant morphisms Y→X where Y is a principal G-space and whose morphisms are commutative diagrams

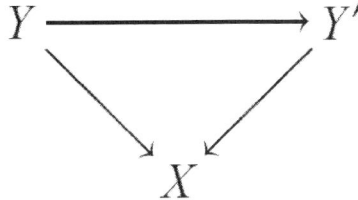

of equivariant morphisms. There is a natural functor $\Phi : \mathrm{Res}_G(X) \to \mathrm{Var}$ to the category of varieties given by Y↦Y/G. Both Sh(−) and D(−) can naturally be viewed as fibred categories over Var. It is shown in [2, Section 2] that the fibres of these over the functor Φ are respectively the category $Sh_G(X)$ of constructible equivariant sheaves and the constructible bounded below equivariant derived category $D_G(X)$. More concretely an object $\mathscr{A}^\bullet$ of the latter is given by an object $\mathscr{A}^\bullet(Y) \in D(Y/G)$ for each $Y \in \mathrm{Res}_G(X)$ and, functorially, for each morphism $\alpha : Y \to Y'$ a quasi-isomorphism $\mathscr{A}^\bullet(Y) \cong \alpha^* \mathscr{A}^\bullet(Y')$ $\mathscr{A}^\bullet$. If X is itself a principal G-space then the category $\mathrm{Res}_G(X)$ has a final object, namely X itself, and we see that, as hoped, $D_G(X) \cong D(X/G)$.

The equivariant derived category inherits the structure of a triangulated category. The assignment $\mathscr{A}^\bullet \mapsto \mathscr{A}^\bullet(G \times X)$ defines a forgetful functor For: $D_G(X) \to D(X)$ and both the usual and perverse t-structures on D(X) can be lifted via the forgetful functor to t-structures on $D_G(X)$. The respective hearts are $Sh_G(X)$ and the equivariant perverse sheaves $Perv_G(X)$. Furthermore the intersection cohomology complex $\mathscr{IC}^\bullet(X)$ has an equivariant lift to a simple object $\mathscr{IC}^\bullet_G(X) \in Perv_G(X)$—see [2, Section 5].

Remark 1.2
1. Note that, owing to the distinction between term-by-term isomorphisms and quasi-isomorphisms of complexes, the equivariant derived category $D_G(X)$ is not generally equivalent to the derived category of $Sh_G(X)$.

2. Generalising the statement that $D_G(X) \cong D(X/G)$ for a principal G-space X, we can interpret the equivariant sheaves and the equivariant derived category as $Sh([X/G])$ and $D([X/G])$ where $[X/G]$ is the quotient stack. This follows immediately from the definition of stacks via fibred categories, see for example [3, 2.2].

The functors $\varphi_*\varphi^*, \varphi_!\varphi^!$ and $\otimes$ extend to the equivariant context (where ϕ is now an equivariant map) simply by defining $(\varphi * \mathscr{A}^\bullet)(Y)$ $\varphi^*(\mathscr{A}^\bullet(Y))=$ etc. There are also two new functors in the equivariant setting. Suppose $\gamma - H \to G$ is a map of reductive algebraic groups and $\varphi : X \to X'$ a map from an H-variety X to a G-variety X' such that $\phi(hx)=\gamma(h)\phi(x)$. Then in Section 6 of [2] Bernstein and Lunts define a functor

$$Q\varphi_* : \mathbf{D}_H(X) \longrightarrow \mathbf{D}_G(X')$$

and a left adjoint

$$Q\varphi^* : \mathbf{D}_G(X') \longrightarrow \mathbf{D}_H(X).$$

Naturally when γ is the identity these agree with $\varphi *$ and φ^*. We define the equivariant intersection cohomology groups of X to be

$$IH_G^*(X) := H^{*+d_X}(Q\pi_* \mathscr{IC}_G^\bullet(X)),$$

where $\pi : X \to pt$ is the map to a point which is considered as a space for the trivial group, and d_X the complex dimension of X. Note that this definition ignores the extra module structure obtained on the graded vector space $IH_G^*(X)$ from considering the point as a G-space.

Remark 1.3

Even for finite dimensional spaces the push-forward $Q\varphi_*$ does not preserve the bounded derived category—this is easily seen because equivariant cohomology can be infinite dimensional—and it is the need to use this functor which forces us to work with bounded below complexes.

There are two important results which we will use. First there is an equivariant version of the famous decomposition theorem [1, 6.2.5]. Suppose $\varphi : X \to X'$ is a proper G-equivariant morphism. Then we have

Theorem 1.4
see Bernstein and Lunts [2, Section 5]. There is a (non-canonical) direct sum decomposition

$$\varphi_* \mathscr{IC}_G^\bullet(X) \cong \bigoplus_\alpha \iota_{\alpha_*} \mathscr{IC}_G^\bullet(V_\alpha; \mathscr{L}_\alpha)[l_\alpha]$$

where $\mathscr{L}_\alpha$ is an irreducible G-equivariant local system on the smooth part of the closed subvariety V_α of X' and $l_\alpha \in \mathbb{Z}$.

Secondly let us suppose that $1 \to N \to G \to H \to 1$ is a short exact sequence of reductive groups, that X is a G-space upon which N acts with only finite stabilisers and that further all the N-orbits are closed and the geometric quotient map $\varphi : X \to X/N$ is affine. Then we have

Theorem 1.5
Bernstein and Lunts [2, 9.1].

1. The functor $Q\varphi_* : D_G(X) \to D_G(X/N)$ preserves both the usual and the perverset-structures so that it restricts to $Q\varphi_* : Sh_G(X) \to Sh_G(X) \to Sh_H(X/N)$ and $Q\varphi_* : Perv_G(X) \to Perv_H(X/N)[d_N]$ where d_N is the complex dimension of N;

2. $Q\varphi_* Q\varphi^* = id$;

3. $Q\varphi_* \mathscr{IC}_G^\bullet(X) \cong \mathscr{IC}_H^\bullet(X /\!/ N)[d_N]..$

   In this paper we relax the conditions that N acts with finite stabilisers and closed orbits and ask what should then replace the third statement above. We find that more generally $\mathscr{IC}^\bullet_H(X /\!/ N)[d_N]$ is a direct summand of $. Q\varphi_* \mathscr{IC}^\bullet_G(X)$

DEFINING A PULL-BACK FOR THE QUOTIENT MAP

Suppose L is an ample line bundle on an irreducible projective variety X upon which a connected reductive algebraic group G acts L-linearly. Suppose N is a connected normal subgroup of G and let H be the quotient:

$1 \to N \to G \to H \to 1$.

We will write X^s and X^{ss} for the subsets of stable and semistable points with respect to the induced N-linearisation on L (not with respect to the G-linearisation). We assume that $X^s \neq \emptyset$. Consider the geometric invariant theory quotient

$$X^{ss} \xrightarrow{\varphi} X^{ss} /\!/ N,$$

where X^{ss} is the Zariski open set of N-semistable points. The normality of N means that X^{ss} is G-invariant. Thus G acts on $X^{ss} /\!/ N$ via the homomorphism $G \to H$. We would like to apply the equivariant decomposition theorem to φ, which is G-equivariant with respect to these actions. However we cannot do so because φ is not, in general, proper. To circumvent this difficulty we consider φ as a rational map from X to $X^{ss} /\!/ N$ and (equivariantly) resolve the points of indeterminacy:

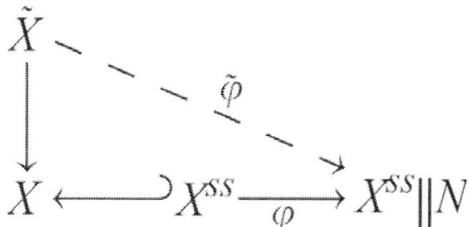

where $\tilde{X}$ is the blowup of X along a closed subscheme supported on X $\setminus X^{ss}$ and $\tilde{\varphi}$ extends φ, see [17, Theorem 1] and cf. [6, 7.17.3]. By the equivariant decomposition theorem we then have a (non-canonical) decomposition

$$\tilde{\varphi}_* \mathscr{IC}_G^\bullet(\tilde{X}) \cong \bigoplus_\alpha \iota_{\alpha_*} \mathscr{IC}_G^\bullet(V_\alpha; \mathscr{L}_\alpha)[l_\alpha]. \tag{2}$$

The Decomposition Theorem and the Intersection Cohomology

Since $\tilde{\varphi}$ is onto there exists a non-empty Zariski open $U \subset X^{ss} \| N$ such that upon applying the forgetful functor we have

$$\tilde{\varphi}_* \mathscr{IC}^\bullet(\tilde{X})|_U \cong \bigoplus \mathscr{L}_\alpha|_U[l_\alpha].$$

As X^s is open and, by assumption, non-empty we may further assume that $U \subset \varphi(X^s)$. Since the fibres of φ are connected $H^{-d_X}(\varphi_* \mathscr{IC}^\bullet(X^{ss})|U)$ is the constant sheaf on U and the restriction

$$H^{-d_X}(\tilde{\varphi}_* \mathscr{IC}^\bullet(\tilde{X})|_U) \to H^{-d_X}(\varphi_* \mathscr{IC}^\bullet(X^{ss})|_U)$$

is onto. It follows that $H^{l_\alpha - d_X}(\mathscr{L}_\alpha|U)$ must also be the constant sheaf for some α. In other words $\mathscr{L}_\alpha|U$ is the constant sheaf in degree $l_\alpha - d_X$ so that

$$i_{\alpha_*} \mathscr{IC}_G^\bullet(V_\alpha; \mathscr{L}_\alpha)[l_\alpha] \cong \mathscr{IC}_G^\bullet(X^{ss} \| N)[d_N]$$

is a direct summand of $\varphi_* \mathscr{IC}^\bullet{}_G(\tilde{X})$.

Let Q_* be the push-forward functor $D_G(X^{ss} \| N) \to D_H(X^{ss} \| N)$. Since N acts trivially on $X^{ss} \| N$

$$Q_* \mathscr{IC}_G^\bullet(X^{ss} \| N) \cong \mathscr{IC}_H^\bullet(X^{ss} \| N; \mathscr{L})$$

where $\mathscr{L}$ is a local system with stalk H_N^*. In particular since $H_N^0 = \mathbb{Q}$ there is a morphism $\mathscr{IC}^\bullet{}_H(X^{ss} \| N) \to Q_* \mathscr{IC}^\bullet{}_G(X^{ss} \| N).\mathscr{IC}^\bullet$. Hence we can define a composition

$$\mathscr{IC}_H^\bullet(X^{ss} \| N)[d_N] \to Q_* \mathscr{IC}_G^\bullet(X^{ss} \| N)[d_N] \to Q\tilde{\varphi}_* \mathscr{IC}_G^\bullet(\tilde{X}) \to Q\varphi_* \mathscr{IC}_G^\bullet(X^{ss}) \quad (3)$$

which we denote by λ.

Remark 2.1

Intuitively we think of this as a pull-back induced by $\varphi : X^{ss} \to X^{ss} \| N$. We must be careful with this viewpoint however because λ is not necessarily unique. The example to bear in mind is that of a variety V which has two small resolutions W_1 and W_2. Let W be a common resolution so that we have:

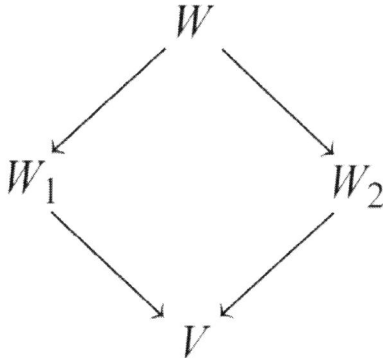

We know that there are natural isomorphisms $H^*(W_1) \cong IH^*(V) \cong H^*(W_2)$ of graded vector spaces but that the ring structures on $H^*(W_1)$ and $H^*(W_2)$ may differ so that their images in $H^*(W)$ cannot be the same. This shows that we cannot expect a canonical pull-back $IH^*(V) \to H^*(W)$.

Corollary 2.2

The morphism λ induces a map $IH^*_H(X^{ss}||N) \to IH^*_G(X^{ss})$ on hypercohomology groups.

Example 2.3

Let us suppose X is a smooth projective variety and take N=G. There are two cases where an explicit description of a subspace of $H^*_G(X^{ss})$ corresponding to $IH^*(X^{ss}||G)$ is already known.

1. First suppose that $G = \mathbb{C}^*$. Let $\mathscr{F}^0$ be the set of fixed point components of $\mathbb{C}^*$ in X^{ss}. For $F \in \mathscr{F}^0$ we define N^+_F to be the set of $x \in X^{ss}$ such that $\lim_{t\to 0} tx \in F$, and similarly N^-_F to be the set of points such that $\lim_{t\to\infty} tx \in F$. Set

    $$c(F) = 2\min(\dim_{\mathbb{C}} N^+_F, \dim_{\mathbb{C}} N^-_F) - 1.$$

Since $\mathbb{C}^*$ acts trivially on F we have $H_{\mathbb{C}^*}^* F \cong H^*(F) \otimes H_{\mathbb{C}^*}^*$. In [11, Section 5] it is shown that $IH^*(X^{ss} \| \mathbb{C}^*)$ is isomorphic (as a vector space) to the set of classes in $H_{\mathbb{C}^*}^*(X^{ss})$ whose restriction to $H_{\mathbb{C}^*}^*(F)$ has degree $<c(F)$ in the second factor of $H^*(F) \otimes H_{\mathbb{C}^*}^*$ for each $F \in \mathcal{F}^0$. Furthermore it is possible to explicitly construct a morphism

$$\mathcal{IC}^\bullet(X^{ss} \| \mathbb{C}^*) \to Q\varphi_* \mathcal{C}_{\mathbb{C}^*}^*(X^{ss})$$

which induces this identification, and to see that it is essentially unique.

2. The second situation is when the action of G (which need not now be $\mathbb{C}^*$) on X is weakly balanced. This notion was introduced in [10]. It consists of two conditions, the first of which ensures that the quotient map $\varphi : X^{ss} \to X^{ss} \| G$ is sufficiently well behaved that there is a natural pull-back φ^* from $IH^*(X^{ss} \| G)$ to $H_G^*(X^{ss})$. In fact φ is, in a slightly extended sense, placid—see [9]. The second part of the weakly balanced condition allows us to identify the image of this pull-back and to show that it is in fact injective. The image is described explicitly in [10].

In both the above examples a certain choice of λ induces an inclusion of the intersection cohomology of the quotient $X^{ss} \| G$ into the G-equivariant cohomology of X^{ss}. This is not a coincidence of these examples but, as we shall see in the next section, a general feature of our situation. What perhaps is special about these examples is that there is a canonical choice of inclusion.

STABLE RESOLUTIONS

We remind the reader that, as above, whenever we write a superscript (s)s we mean (semi)stability with respect to the normal subgroup N of G. Let us fix a choice of morphism

$$\lambda : \mathcal{IC}_H^\bullet(X^{ss} \| N)[d_N] \to Q\varphi_* \mathcal{IC}_G^\bullet(X^{ss})$$

as above. We now show how to construct a morphism

$$\kappa : Q\varphi_* \mathscr{I}\mathscr{C}_G^\bullet(X^{ss}) \longrightarrow \mathscr{I}\mathscr{C}_H^\bullet(X^{ss} /\!/ N)[d_N]$$

such that $\kappa\lambda =$ id. More precisely we show that every N-stable resolution of (X, X^{ss}) induces such a morphism.

Definition 3.1

An N-stable resolution of (X, X^{ss}) is given by a commutative diagram of G-spaces

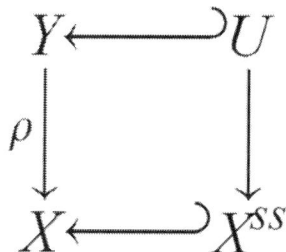

where ρ is proper and birational, and U an open subset of Y. Further we assume that there is an ample line bundle $M \rightarrow Y$ and a G-linearisation on M such that

1. every point of U is N-stable for $M|_U$ (but we do not necessarily assume that $U \subset Y^s$);

2. the induced map $\sigma : U|\!|N \rightarrow X^{ss}|\!|N$ is proper and birational.

We will usually suppress M and simply write $\rho : (Y, U) \rightarrow (X, X^{ss})$ or even just ρ. Note that our assumption that $X^s \neq \emptyset$ implies that there is a non-empty open subset V of X^s such that the restriction of ρ to $\rho^{-1}V \rightarrow V$ is an isomorphism.

Example 3.2

1 Suppose we can choose a new linearisation of the G action on X with the properties that N-stability and semistability coincide i.e. $X^{ss}_{new} = X^s_{new}$ and that $X^s_{new} \subset X^{ss}$. Then the inclusion

$$(X, X^s_{new}) \hookrightarrow (X, X^{ss})$$

is an N-stable resolution.

2　We reinterpret the paper [15] in the framework we have introduced. (In fact [15] is more general since it applies to symplectic reductions by S^1 and not just to algebraic quotients.)

Suppose X is smooth and $G=N=\mathbb{C}^*$. Let $\mathscr{F}^0$ be the set of fixed point components of $\mathbb{C}^*$ in X^{ss}. For $F \in \mathscr{F}^0$ we define N_F^+ to be the set of x $\in X^{ss}$ such that $\lim_{t\to 0} tx \in F$, and similarly N_F^- to be the set of points such that $\lim_{t\to\infty} tx \in F$. Then

$$X^s = X^{ss} \setminus \bigcup_{F \in \mathscr{F}^0} N_F^+ \cup N_F^-.$$

We define a stable resolution by taking Y=X (with ρ =id) and

$$U = X^{ss} \setminus \bigcup_{F \in \mathscr{F}^0} N_F^{d(F)},$$

where $d(F)=\pm$ depending upon whether $\dim_{\mathbb{C}} N_F^+$ or $\dim_{\mathbb{C}} N_F^-$ is the larger, if they are equal then we make an arbitrary choice. (Note that in this case $U \not\subset X^s$ but that every point of U is nevertheless stable for the restriction of L to U.) The induced map $U||\mathbb{C}^* \to X^{ss}||\mathbb{C}^*$ is shown to be a small resolution in [15]. This follows from an application of the ideas of [7] to a G-invariant neighbourhood of each $F \in \mathscr{F}^0$.

Hence $IH^*(X^{ss}||\mathbb{C}^*) \cong IH^*(U||\mathbb{C}^*) \cong H^*_{\mathbb{C}}(U)$. Lerman and Tolman then go on to compute the kernel of the restriction $H^*_{\mathbb{C}^*}(X^{ss}) \to H^*_{\mathbb{C}^*}(U)$, and so express $IH^*(X^{ss}||\mathbb{C}^*)$ as a quotient of $H^*_{\mathbb{C}^*}(X^{ss})$.

3　Suppose X is smooth and take N=G. In [13] Kirwan describes a canonical partial desingularisation of the quotient $X^{ss}||G$. This is constructed by taking the quotient of a projective variety Y (with a suitably linearised G action) obtained from X by a sequence of blowups. Initially we blow up along the closure in X of the smooth G-invariant subvariety

$$GZ_R^{ss} = \{x \in X^{ss} \mid \operatorname{Stab}_G x \text{ conjugate to } R\}$$

where R is a connected reductive subgroup of G such that GZ_R^{ss} has maximal codimension amongst all such subvarieties. There is an induced action of G on the blowup $\pi : \tilde{X} \to X$ which can be linearised on the ample bundle formed by twisting the pull-back of a sufficiently large power of L with minus the exceptional divisor. It is shown in [13, Lemma 6.1] that $\pi : (\tilde{X}^{ss}) \subset X^{ss}$ and $G\tilde{Z}_R^{ss} = \varnothing$. Continuing inductively we can construct a G-stable resolution.

We can refine this procedure slightly by working relative to a nontrivial normal subgroup N. Now we consider blowing up along the closures of subvarieties of the form GZ_R^{ss} where ss refers to N-semistability and R is a connected reductive subgroup of N. The same procedure will then construct an N-stable resolution.

Proposition 3.3
There is always at least one N-stable resolution.

Proof
To construct a resolution we first of all G-equivariantly resolve the singularities of X (see [17, Theorem 1]). Thus we have a smooth G-variety $\tilde{X}$ and a G-equivariant map $\tilde{X} \to X$ which factors as a finite sequence of blowups of smooth G-invariant subvarieties.

Suppose the first of these blowups is $\pi : \tilde{X} \to X$. The action of G on $\tilde{X}$ can be linearised on an ample line bundle of the form $\pi^* L^d \otimes \mathcal{O}(-E)$ where E is the exceptional divisor and d is sufficiently large. As usual let $\hat{X}^{ss}$ denote the N-semistable points of $\tilde{X}$ with respect to this linearisation. I claim that $\pi(\hat{X}^{ss}) \subset X^{ss}$. To see this note that if $x \in \hat{X}^{ss}$ then there is a N-invariant section $\sigma = \sigma_1 \otimes \sigma_2$ of $\pi^* L^d \otimes \mathcal{O}(-E)$ with $\sigma(x) \neq 0$. Since $\mathcal{O}(-E)$ is trivial away from E we deduce that σ_1 is N-invariant and, of course, $\sigma_1(x) \neq 0$. Equivalently $\pi(x)$ is N-semistable for the obvious linearisation on Ld. But by the final remark of [16, Chapter 1, Section 5] the N-semistable points for the linearisations on L and L^d coincide.

Proceeding inductively we see that we can linearise the action of G on $\tilde{X}$ in such a way that we have a map $(\tilde{X}, \tilde{X}^{ss}) \to (X, X^{ss})$. We can now apply Kirwan's resolution, relative to N, to the smooth G-variety $\tilde{X}$ (see Example 3 of 3.2). In this way we will obtain an N-stable resolution of (X, X^{ss}).

Let ρ be an N-stable resolution of X^{ss} so that, in the notation of 3.1, we have a diagram:

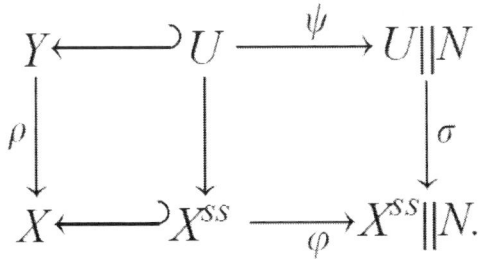

We can apply the equivariant decomposition theorem to the proper maps ρ and σ to obtain morphisms

$$\mathscr{IC}_G^\bullet(X^{ss}) \to \rho_* \mathscr{IC}_G^\bullet(U) \quad \text{and} \quad \sigma_* \mathscr{IC}_H^\bullet(U/\!/N) \to \mathscr{IC}_H^\bullet(X^{ss}/\!/N).$$

These are of course not canonical but their restrictions to V and $\varphi(V)$ respectively are the natural quasi-isomorphisms induced by ρ and ρ^{-1}. By Theorem 1.5 $Q\psi_* \mathscr{IC}_G^\bullet(U) \cong \mathscr{IC}_H^\bullet(U/\!/N)[d_N]$ because N acts with only finite stabilisers on U. So we can compose these morphisms to obtain

$$\kappa : Q\varphi_* \mathscr{IC}_G^\bullet(X^{ss}) \to \mathscr{IC}_H^\bullet(X^{ss}/\!/N)[d_N]. \tag{4}$$

Theorem 3.4
The composition $k\lambda\,[-d_N]$ is the identity on $\mathscr{IC}_H^\bullet(X^{ss}/\!/N)$.

Proof

We know from [2, Section 5] that $\mathscr{IC}^{\bullet}_{H}(X^{ss}||N)$ is a simple object in the heart of the perverse t-structure on $D_{H}(X^{ss}||N)$ so that

$$\mathrm{Hom}(\mathscr{IC}^{\bullet}_{H}(X^{ss}||N), \mathscr{IC}^{\bullet}_{H}(X^{ss}||N)) \cong \mathbb{Q}.$$

We can easily check that $k\lambda\,[-d_{N}]$ restricts to the identity on $\varphi(V)$ and hence must be the identity.

Corollary 3.5

The map $IH^{*}_{H}(X^{ss}||N) \hookrightarrow IH^{*}_{G}(X^{ss})$ induced by λ is an inclusion. Any N-stable resolution can be used to define (not necessarily uniquely) a projection $IH^{*}_{G}(X^{ss}) \to IH^{*}_{H}(X^{ss}||N)$ which is split by this inclusion.

Proof

This follows immediately from Proposition 3.3 and Theorem 3.4.

Remark 3.6

In [14] the partial desingularisation constructed in [13] (see example 3.2 part 3) is used to define, as we have above, a map $H^{*}_{G}(X^{ss}) \to IH^{*}(X^{ss}||G)$ when X is smooth. [14, Theorem 2.5] states that this map is a surjection. This is then used to give an algorithm for computing the intersection Betti numbers of $X^{ss}||G$.

Unfortunately the given argument that the map is surjective is incorrect due to a confusion on page 499 where it says "by 2.24 there is a commutative diagram

$$
\begin{array}{ccc}
H^{*}_{G}(E)/H^{*}_{N}(Z^{ss}_{R}) & \longrightarrow & H^{*}_{G}(Y) \\
\downarrow & & \downarrow \hat{\phi}_{Y} \\
IH^{*}(E||G)/IH^{*}(\mathscr{N}/G) & \longrightarrow & IH^{*}(Y||G)
\end{array}
$$

where the horizontal maps are the inclusions associated to these decompositions". In fact Kirwan's 2.24 shows that the same four objects form a commutative diagram but where the horizontal maps arise from Gysin sequences and are not those associated to "these decompositions".

Corollary 3.5 provides a new proof that Kirwan's map is surjective and hence re-validates the algorithm.

ACKNOWLEDGMENTS

I am greatly indebted to interesting and illuminating conversations with both Young-Hoon Kiem and Sue Tolman. I would also like to thank Frances Kirwan for her patience in reading preliminary drafts.

REFERENCES

1. A. Beilinson, J. Bernstein, P. Deligne, Faisceaux pervers, Astérisque 100 (1982). Proceedings of the C.I.R.M. Conférence, Analyse et topologie sur les espaces singuliers.
2. J. Bernstein, V. Lunts, Equivariant sheaves and functors, in: Lecture Notes in Mathematics, Vol. 1578, Springer-Verlag, Berlin, 1994.
3. T. GXomez, Algebraic stacks, Proc. Indian Acad. Sci. Math. Sci. 111 (1) (2001) 1–31.
4. M. Goresky, R. Macpherson, Intersection homology theory II, Invent. Math. 71 (1983) 77–129.
5. V. Guillemin, J. Kalkman, The JeGrey–Kirwan localization theorem and residue operations in equivariant cohomology, J. Reine Angew. Math. 470 (1996) 123–142.
6. R. Hartshorne, Algebraic Geometry, Springer-Verlag, Berlin, 1977.
7. Y. Hu, The geometry and topology of quotient varieties of torus actions, Duke Math. J. 68 (1) (1992) 151–184.
8. L. JeGrey, F. Kirwan, Localization for non-abelian groups, Topology 34 (2) (1995) 291–327.
9. Y.-H. Kiem, Note on placid maps, preprint, Math. Dept. Yale, October 2000.
10. Y.-H. Kiem, Intersection cohomology of quotients of nonsingular varieties, Invent. Math. (2002), submitted for publication.
11. Y.-H. Kiem, J. Woolf, The cosupport axiom, equivariant cohomology and the intersection cohomology of certain symplectic quotients, Can. J. Math. (2002), submitted for publication.

12. F. Kirwan, Cohomology of Quotients in Symplectic and Algebraic Geometry, in: Mathematical Notes, Vol. 34, Princeton University Press, Princeton, 1984.
13. F. Kirwan, Partial desingularisations of quotients of nonsingular varieties and their Betti numbers, Ann. Math. 122 (1985) 41–85.
14. F. Kirwan, Rational intersection homology of quotient varieties, Invent. Math. 86 (1986) 471–505.
15. E. Lerman, S. Tolman, Intersection cohomology of S1 symplectic quotients and small resolutions, Duke Math J. 103 (1) (2000) 79–99.
16. D. Mumford, J. Fogarty, F. Kirwan, Geometric Invariant Theory, Third Edition, Springer-Verlag, Berlin, 1994.
17. Z. Reichstein, B. Youssin, Equivariant resolution of points of indeterminacy, Proc. Amer. Math. Soc. 130 (8) (2002) 2183–2187.
18. S. Tolman, J. Weitsman, On the cohomology rings of Hamiltonian t-spaces, Northern California Symplectic Geometry Seminar, 1999, pp. 251–258, Amer. Math. Soc. Transl. Ser. 2, 196.
19. E. Witten, Two dimensional gauge theories revisited, J. Geom. Phys. 9 (1992) 303–368.

CITATION

Jonathan Woolf, The decomposition theorem and the intersection cohomology of quotients in algebraic geometry, Journal of Pure and Applied Algebra, Volume 182, Issues 2–3, 1 August 2003, Pages 317-328, ISSN 0022-4049, http://dx.doi.org/10.1016/S0022-4049(03)00030-6.

Topics on Phylogenetic Algebraic Geometry

Cristiano Bocci

Dipartimento di Matematica, Università di
Milano, Via Cesare Saldini 50,
20133 Milano, Italy

ABSTRACT

This expository work deals with the main aspects of Phylogenetic Algebraic Geometry. In particular, we will focus our attention on the technique of flattening of a n-dimensional tensor. Our interest in flattening is due to the fact that this is strictly related to the study of secant varieties of Segre varieties.

MSC

INTRODUCTION

One of the main problems in modern Biology is that of phylogenetic inference. Let us consider a model of molecular evolution (for example, DNA sequences) and suppose that evolution occurs along a bifurcating tree, proceeding from a root, i.e. the common ancestral species, toward the leaves, i.e. the descendant species. We require that, at each site in the sequences, bases mutate according to a probabilistic process that depends upon the edges of the tree. Only the sequences at the leaves of the tree can be observed, while sequences at internal nodes correspond to hidden variables in this graphical model. Thus, the phylogenetic inference concerns the problem to infer the tree topology from observed sequences, assuming some probabilistic (and reasonable) model.

In 1987, Cavender and Felsenstein [7] and, separately, Lake [22], introduced an algebraic approach to attack this problem. In fact, under many standard models of molecular evolution, for a fixed tree topology, the joint distribution of bases at the leaves are described by polynomial equations in the parameters of the model. They proposed to search for polynomials, called phylogenetic invariants, which vanish on any joint distribution arising from the tree and model, regardless of parameter values.

Recently, several authors have started to research and study phylogenetic invariants by a deeper use of Algebraic Geometry. Although the idea of Phylogenetic Algebraic Geometry had already been undertaken in their works (for example, [1], [2], [3], [24] and [28]) only in [12] we can find its definition for the first time. Here the authors say that Phylogenetic Algebraic Geometry is concerned with certain complex projective algebraic varieties derived from finite trees. By Phylogenetic Algebraic Geometry we mean the study of algebraic varieties which represent statistical models of evolution.

The varieties which arise from such a kind of model can be different. We can find, for example, the more familiar ones, as secant varieties, determinantal varieties, toric varieties and Segre–Veronese varieties. This happens, in general, when we consider models related to small trees, i.e. trees with at most five leaves. For trees with more than six leaves instead, we can encounter families of new kinds of varieties, often completely unknown. The study of such varieties is related especially to the search for the generators of their ideals. By the Hilbert Basis Theorem we know that these generators are in a finite number and are exactly the phylogenetic invariants associated to the corresponding tree.

These invariants are defined rigorously in Section 3. Here, we introduced also the basic facts on phylogenetic invariants which, in Section 4, are analyzed from an Algebraic Geometry viewpoint. After the definitions of phylogenetic ideal and phylogenetic variety, we briefly describe some important results. It is surely impossible to cite here all the recent issues on Phylogenetic Algebraic Geometry. Thus, we prefer to focus our attention on the Flattening technique, which is explained in Section 5. This technique permits to introduce a fundamental topic in

Algebraic Geometry, namely the secant variety. In Section 6, we show why secant varieties are so useful to search for phylogenetic invariants. It is important to notice that several recent papers on Algebraic Geometry (like [6] and [23]) are really helpful to such research.

For general background reading on Phylogenetics, we strongly suggest the books by Felsenstein [14] and Semple-Steel [26]. They deeply analyze evolutionary trees according to Biology, Computer Science, Statistics and Mathematics. Instead, for a survey of Algebraic Statistics and Computational Biology, the book [25], edited by Lior Pachter and Bernd Sturmfels, is surely the best choice. There is large literature about Algebraic Geometry and Commutative Algebra. The elements we need, though, can be found in [10](Chapter 0), [18] (Lectures 1, 2 and 8) and [19] (Chapter 1). For the interested reader, we suggest also books [8] and [9], where the authors introduce concepts and results in Algebra and Geometry with the perspective of possible applications. For references to others research papers we recommend again [14] and [26]. For the most recent ones, the reader can consult [12]. Here, there is also a very interesting section where the authors collect a series of open problems.

We would like to mention that our paper follows closely the works of E. S. Allman and J. A. Rhodes ([1], [2],[3] and [4]), especially as far as notation and development of ideas are concerned.

We write Z, R, and C, respectively, for the ring of integers and the fields of real and complex numbers.

EVOLUTIONARY TREES AND MARKOV MODELS

Since graphs play an important role in phylogenetics, we will start recalling some basic facts about it.

Definition 2.1: A graph G is an ordered pair (V, E) consisting of a non-empty set V of vertices and a multiset E of edges each of which is an element of $\{(x,y): x,y \in V\}$. An edge that joins a vertex to itself is a loop and the edges that join the same distinct pair of vertices are called parallel edges.

All the graphs we consider will have a finite set of vertices.

If $e = \{u, v\}$ is an edge of a graph G, then u and v are adjacent and e is said to be incident with u and v. The vertices u and v are the ends of e. Let v be a vertex of a graph G. The valency of v, $v(v)$, is the number of edges in G that are incident with v. A path in a graph G is a sequence of distinct vertices $v_1, v_2, \ldots, v_k$ such that, for all $i = 1, \ldots, k-1$, v_i and v_{i+1} are adjacent. If, in addition, v_1 and v_k are adjacent, then the subgraph of G, whose vertex set is $\{v_1, v_2, \ldots, v_k\}$ and whose edge set is $\{(v_k, v_1)\} \cup \{(v_i, v_{i+1}) : i = 1, \ldots, k-1\}$, is a cycle. A graph is connected if each pair of vertices in G can be joined by a path: otherwise G is disconnected (Fig. 1).

Let us denote by $|F|$ the cardinality of a set F. We recall the following

Lemma 2.2: Let $G = (V, E)$ be a graph, then

$$\sum_{v \in V} v(v) = 2|E|.$$

Moreover, if G is connected, one has $|V| \le |E| + 1$.

Graphs have several applications in Biology: food web and competition graphs, genome mapping andinterval graphs, pedigree (di) graphs. Here, we deal with another application: the theory of phylogenetic trees.

Definition 2.3: A tree $T = (V, E)$ is a connected graph with no cycles. A tree is a path graph if all vertices have valency at most two.

An important characterization of trees is given by

Theorem 2.4: Let $G = (V, E)$ be a graph. Then the following are equivalent:

1) G is a tree;
2) For any two vertices v and u in V there exists a unique path in G from v to u;
3) G is connected and $|V| = |E| + 1$.

Topics on Phylogenetic Algebraic Geometry

a   b

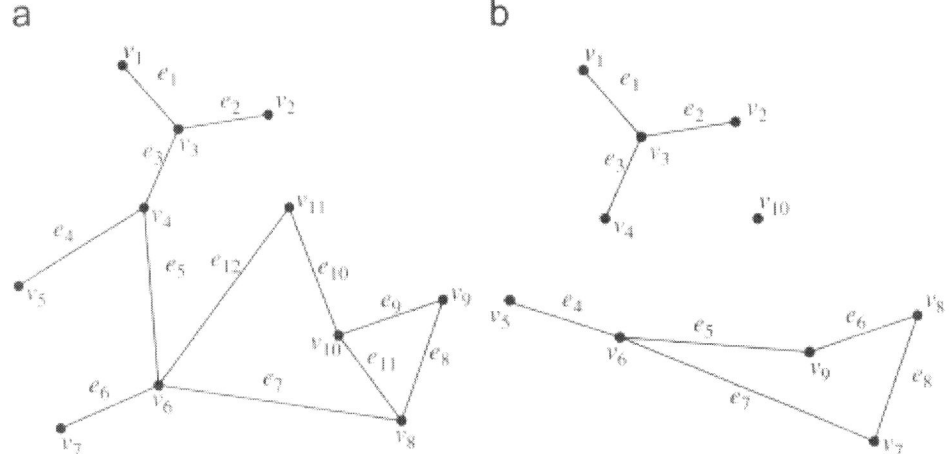

Figure.1: (a) A connected graph and (b) a disconnected graph.

A vertex of a tree of valency one is called a leaf. We denote by L the set of leaves and define $\tilde{V} := V \setminus L$ the set of interior vertices. Similarly, we denote by $\tilde{E}$ the set of interior edges. A tree is binary, or bifurcated, if every interior vertex has valency three. Two distinct leaves of a tree are said to form a cherry if they are adjacent to a common vertex. For example, in Figure.2, the pairs $\{v_1, v_2\}$ and $\{v_7, v_8\}$ are cherries.

A rooted tree is a tree that has exactly one distinguished vertex called the root, which we denote by the letter r. For a rooted tree T we can define a natural partial order $\leq_T$ on the vertex set V by

$v_i \leq_T v_j$ if the path from the root of T to v_i includes v_j.

In this case we say that v_j is a descendant of v_i and that v_i is an ancestor of v_j. For this reason we always draw a rooted tree with the root r at the top of the figure and oriented so as to respect the ancestor-descendant relationship (Figure.3).

Let us state, now, the definition of phylogenetic tree. Among several definitions, we will choose the one closer to Biology:

Definition 2.5: An X-tree T is an ordered pair (T, ϕ), where T is a tree with vertex set V, label set X and $\phi : X \to V$ is a map with the property that, for each $v \in V$ of valency at most two, $v \in \varphi(X)$. An X-tree is also called a semi-labelled tree (on X). A phylogenetic tree is an X-tree (T, φ) with the property that φ is a bijection from Xinto the set of leaves of T. If, in addition, every interior vertex of T has valency three, T is a binary phylogenetic tree.

It is common in Biology to focus on binary trees (i.e., trivalent, except bivalent at the root) as being of primary interest. In fact, most speciation events are believed to be of the sort where only two species at a time arise from a parent species. Multifurcations in a tree might be used to represent ignorance such as when several speciation events occur so closely in time that we are unable to resolve their order.

From now on, if not otherwise specified, we will consider only binary trees.

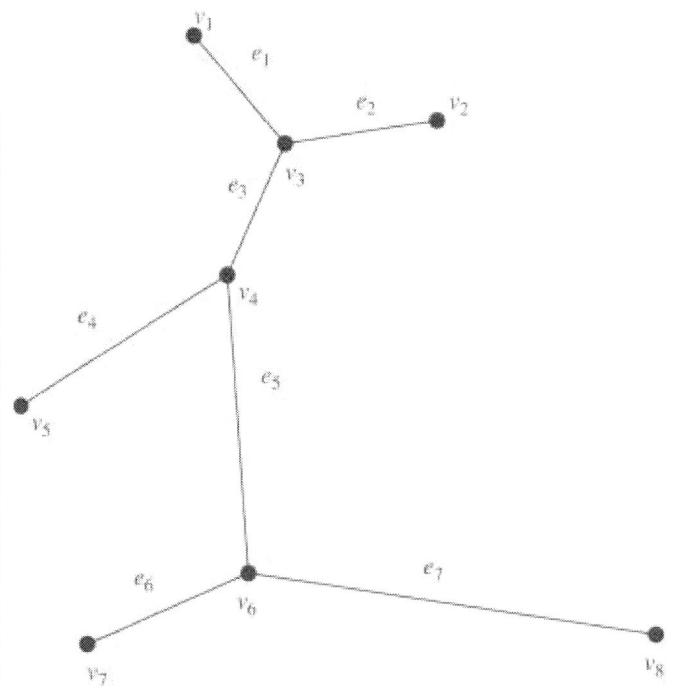

Figure.2: An example of tree.

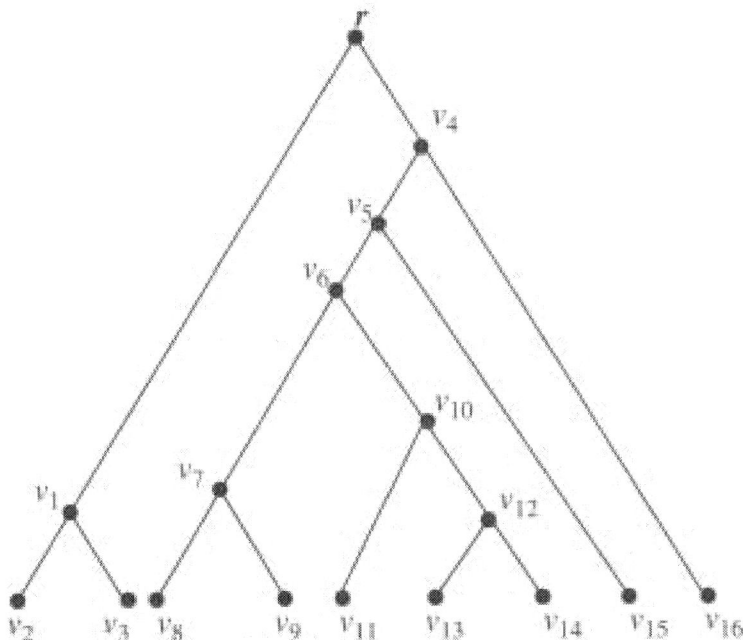

Figure.3: A rooted tree.

Proposition 2.6: (i) Let T be a binary phylogenetic X-tree and let $n = |X|$, Then, for all $n \geq 2$, T has $2n-3$ edges and $n-3$ interior edges.

(ii) Let $B(n)$ be the set of all binary phylogenetic trees with label set $X = \{1, 2, \ldots, n\}$. If $n = 2$ then $|B(n)| = 1$. If $n \geq 3$ then

$$|B(n)| = \frac{(2n-4)!}{(n-2)!2^{n-2}} = 1 \times 3 \times 5 \times \cdots \times (2n-5).$$

Proof: See [26, Propositions 2.1.3 and 2.1.4].

Obviously, we can extend the notion of an X-tree to the rooted case.

Definition 2.7: A rooted X-tree T is an ordered pair (T, ϕ), where T is a rooted tree with vertex set V, rooted vertex r, label set X and $\phi : X \rightarrow V$ is a map with the property that, for each $v \in V \setminus \{r\}$ of valency at most two, $v \in \phi(X)$. A rooted X-tree is also called a rooted semi-labelled tree (on X). A rooted phylogenetic tree is a rooted X-tree (T, ϕ) with the property that ϕ is a bijection from X into the set of leaves of T and the root has valency at least two. If, in addition, every interior vertex of T has valency three, T is a rooted binary phylogenetic tree.

Let T be a rooted X-tree and let x,y be two leaves. We denote $lca(x,y)$ the most recent common ancestor of x and y. For example, in Figure.3, one has $lca(_{v8,v9})=_{v7}$, $lca(_{v8,v11})=lca(_{v8,v12})=_{v6}$, $lca(_{v8,v16})=_{v4}$. For a rooted phylogenetic tree T on X, we view the edges of T as being directed from the root r. Then we consider T as describing the evolution of the set X of extant species that label the leaves of T from a common hypothetical ancestral species at r; the other interior vertices of T correspond to further hypothetical ancestral species or to past speciation events. Thus $lca(x,y)$ can be seen as the most recent shared ancestral species (or speciation event) of the species x and y.

Remark 2.8: Unrooted phylogenetic trees are also biologically relevant because they are typically what the tree reconstruction methods generate. We can observe that it is always possible to pass from an unrooted tree to a rooted one and viceversa. In particular, passing from the unrooted to the rooted tree, means to choose an internal vertex as the root or add another vertex inside an edge and choose it as the root.

Remark 2.9: In general, as X, we will use the set $\{1,2,...,n\}$, where each number will correspond to a specific species.

Remark 2.10: Let us mention some particular kinds of trees. The (rooted) caterpillar tree is any (rooted) binary phylogenetic tree for which the induced subtree on the interior vertices is a path graph. A rooted balanced tree of height $h \geq 0$ is a rooted binary phylogenetic tree, with $n=2^h$ leaves, each of which is separated from the root by exactly h edges. A star tree is a phylogenetic tree with no interior edges, i.e. with a single interior vertex that is adjacent to all the leaves. In Figs. 4 and 5 we show, respectively, the unrooted and rooted cases. Each rooted case is obtained by adding another vertex inside an edge in the respective unrooted case.

We introduce now the concept of Markov process.

Let $x_1,...,x_t$ be random variables on a sample space S taking value in a set U and let $A=\{1,2,...,t\}$. For a subset $B \subset A$ and an event E of

S we will write $\text{Prob}(E|_{\cap_{i\in B}\{x_i\}})$ for $\text{Prob}(E|_{\cap_{i\in B}\{x_i=u_i\}})$, i.e. the conditional probability of E given $_{\cap_{i\in B}}\{x_i=u_i\}$, for every selection of $u_i\in U$.

We fix an alphabet with k letters, for instance $[k] = \{1, 2,\ldots, k\}$.

Definition 2.11: Let T be a rooted tree with vertex set V. A Markov process on T, with state set $[k]$, is a family $\{x_v : v\in V\}$ of random variables such that, whenever (u,v) is arc of T, with $u<v$, and $\alpha\in[k]$,

$$\text{Prob}(x_v=\alpha|\cap_{w<v}x_w)=\text{Prob}(x_v=\alpha|x_u) \tag{2.1}$$

Condition (2.1) is known as the Markov Property. Intuitevely, this states that, for each arc (u,v) of T, the value of x_v, conditional on x_u, is independent of the X-values at all other "earlier" vertices.

Let T be a rooted tree. For each edge $e=(u,v)$ of T (with $u<v$), a Markov process on T, with state set$[k]$, induces an associated $k\times k$ transition matrix, denoted $M^{(e)}$, where the (i,j)-entry, $m_{ij}^{(e)}$, is the probability to pass from state i on u to state j on v. We ask for

(i) $m_{ij}^{(e)} \geq 0$

(ii) $\sum_{j=1}^{k} m_{ij}^{(e)} = 1$.

Thus, if we specify a Markov matrix for each edge of the tree, we have modeled how the entire evolutionary process proceeds along the tree.

Once we fixed the root r we define also a root distribution $\pi(r)=(\pi(r)_1,\pi(r)_2,\ldots,\pi(r)_k)$, where $\pi(r)_i$ is the probability to have the state i at the root. Obviously $\pi(r)_i \geq 0, \forall i=1,\ldots,k$ and $\sum_{j=1}^{k}\pi(r)_i = 1$.

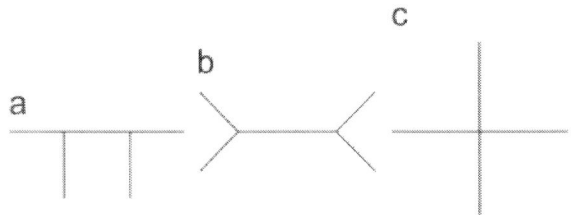

Figure.4: (a)–(c) A caterpillar tree, a balanced tree of height 2, a star tree.

a b c

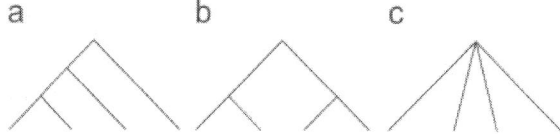

Figure.5: (a)–(c): The rooted cases.

The root distribution vector $\pi(r)$ gives probabilities of the various states for the variable at the root, while $k \times k$ Markov matrices give transition probabilities of state changes from ancestral to descendant nodes along each edge.

Definition 2.12: We refer to data (T, M) as the general Markov model on T, where $M=(\pi(r),\{M^{(e)}:e\in E\})$. We often refer to M as stochastic parameters, distinguishing them from tree parameters.

Remark 2.13: As to the DNA, the number of states is $k=4$, but as to protein sequences, which are built from 20 amino acids, $k=20$. The case $k=2$ is also of interest for DNA substitution models, if we group bases into purines $R=\{A, G\}$ and pyrimidines $Y=\{C, T\}$.

Let $l_1,\ldots,l_n$ be the leaves of the tree T. Evolution occurs along the tree, but we can observe sequences only at the leaves of T. With the parameters of the model M thus specified, we are interested in the joint distribution P of states at the leaves l_i's. The joint distribution P is an n-dimensional $k \times k \times \cdots \times k$ tensor (or table or array) with entries

$$P(_{i1},\ldots,_{in})=Prob(_{l1}=_{i1},\ldots,_{ln}=_{in})$$

where $Prob(_{l1}=_{i1},\ldots,_{ln}=_{in})$ represents the probability to have state $_{ij}$ at the leaf $_{lj}$, for $j=1,\ldots,n$. In general, we will denote P $(_{i1},\ldots,_{in})$ by $p_{i1\ldots in}$. The entries of P are the expected frequencies to be seen in patterns of states $(_{i1},\ldots,_{in})$ at the leaves of the tree. These expected pattern frequencies can be explicitly expressed in terms of the parameters of the model, as we can explain through an example.

Example 2.14: Consider the tree with leaves $l_1, \ldots, l_5$.

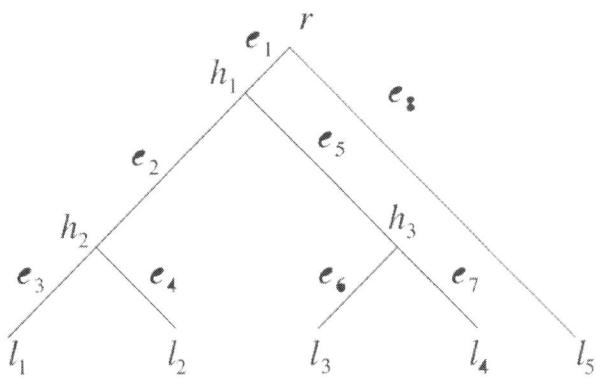

Let $M^{(e)} = (m^{(e)})_{ij}$ be the $k \times k$ matrix on the edge h_e, for $e = 1, \ldots, 8$, and $\pi(r)$ the root distribution. Suppose that we want to compute the probability $p_{i1 \cdots i5}$. If we start from a state $w0$ at the root, we can see that $\pi(r)_{w0} m^{(1)}_{w0,w1}$ is the probability to have a state $w1$ at the vertex h_1. Moving in this way, we can see that we reach leaf l_1 with state $i1$ by

$$\pi(r)_{w0} m^{(1)}_{w0,w1} m^{(2)}_{w1,w2} m^{(3)}_{w2,i1},$$

where $w2$ is an unobserved state at the vertex h_2. The procedure for the leaf l_2 is similar, but, since we already have the probability of transition until vertex h_2, from the computation on l_1, it is enough to multiply the previous term by $m^{(4)}_{w2,i_2}$. Now it is clear how to proceed. Thus we obtain

$$\pi(r)_{w0} m^{(1)}_{w0,w1} m^{(2)}_{w1,w2} m^{(3)}_{w2,i_1} m^{(4)}_{w2,i_2} m^{(5)}_{w1,w3} m^{(6)}_{w3,i_3} m^{(7)}_{w3,i4} m^{(8)}_{w0,i_5}$$

This is the probability to have state ij at the leaf lj, $j = 1, \ldots, 5$, and states $w0', w1', w2$ and $w3'$, respectively, at the root r and at vertices $h1', h2', h3$. Since the internal nodes are hidden, we have to consider all possible states at the internal nodes, thus the final probability will be

$$P_{i_1 \ldots i_5} \quad \sum_{\substack{1 \le w_i \le k \\ i=0,1,2,3}} \pi(r)_{w0} m^{(1)}_{w0,w1} m^{(2)}_{w1,w2} m^{(3)}_{w2,i_1} m^{(4)}_{w2,i_2} m^{(5)}_{w1,w3} m^{(6)}_{w3,i_3} m^{(7)}_{w3,i4} m^{(8)}_{w0,i_5}$$

In general, let $T=(V,E)$ be an n-taxon tree with Markov model $M=(\pi(r), M^{(e)})$. Let us denote by $s(e)$ and $f(e)$ the ends of e. Thus, the joint distribution P is given by the formula

$$P(i_1, \ldots, i_n) = \sum_{(b_v) \in H} \left[\pi(r)_{b_r} \prod_e (m^{(e)}_{b_{s(e)}, b_{f(e)}}) \right],$$

$$(2.2)$$

Where the product is taken over all edges e getting away from the root r and the sum is taken over the set

$$H = \{(_{bv)v \in V} \mid _{bv} \in [k] \text{ if } v \neq _{ij}, b_v = _{ij} \text{ if } v = _{ij}\} \subset [k]^{2n-2}.$$

We can say that H represents the set of all "histories" consistent with the specified states at the leaves. More generally, for the general Markov model on an n-taxon tree, each probability $_{pi1,\ldots,in}$ will be a degree $2n-2$ polynomial, with k^{n-2} terms. The precise form of these polynomials reflects the topology of the tree T.

Remark 2.15: The model we have described here concerns a base substitution process at a single site. In general, for phylogenetic inference, the data are aligned DNA sequences of some length L. Thus, we assume that the evolutionary process at each site proceeds independently of all other sites, but according to the same probabilistic process, with the same parameters. This independent, identically distributed (i.i.d.) assumption is not desirable from a biological viewpoint. In fact, we can have substitutions at one site which are not independent. However, a form of the i.i.d. assumption is essential since only by viewing each site as a trial of the same process, we can obtain enough data to infer something about the parameters.

Let us point out the following important

Proposition 2.16: Fix an n-taxon tree T. Let r be a choice of root for T (which may be a leaf, an internal node of valency 3, or along some edge). Then, for a generic choice of stochastic parameters $_{sr}$ for the general Markov model rooted at r, and for any other choice of a root q for T, on either a leaf or an internal node of valency 3, there is a uniquely determined choice of general Markov model parameters $_{Sq}$ for the model rooted at q producing the same joint distribution at the leaves as $_{Sr}$.

Proof: See [1, Proposition 1].

A consequence of the previous Proposition is that the location of the root in a tree T is a biological problem, not a mathematical one.

The general Markov model has more parameters than models typically use in practice. Once the tree parameter has been chosen as a particular n-taxon tree, there are k-1 free choices on $\pi(r)$ (since $\sum_{j=1}^{k} \pi(r)_i = 1$) and k(k-1) free choices on the entries of the matrix $M^{(e)}$, for each edge e . Thus, one has N:= (2n-3) k (k-1) +k-1 numerical parameters. The growth is only linear in the number of taxa, but the coefficient, depending on k , could be very large. For example, for k=2, the total number of parameters is 4n-2, while, for k=4, it grows as 24n. The number of parameters has several effects on the inference: slow computations, overfitting. If the data can be described by a model with fewer parameters, that model may provide a better basis for inference. Thus, in general, we consider particular restrictions on the stochastic parameters of the general Markov model, given by mathematical and/ or biological reasons.

Let us introduce now some examples of submodels of the general Markov model. The reader can find a wider range of submodels in [4] (the General Time Reversible model and Mixture model) and in [2] (theStable Base Distribution model, the Simultaneous Diagonalization model, the Algebraic Time Reversiblemodel).

The Jukes–Cantor Model for DNA
This model is the biologically plausible model with the fewest parameters. It assumes a uniform root distribution vector $\pi(r) = (0.25, 0.25, 0.25, 0.25)$ and edge transition matrices of the form

$$M^{(e)} = \begin{pmatrix} 1 - a_e & \dfrac{a_e}{3} & \dfrac{a_e}{3} & \dfrac{a_e}{3} \\ \dfrac{a_e}{3} & 1 - a_e & \dfrac{a_e}{3} & \dfrac{a_e}{3} \\ \dfrac{a_e}{3} & \dfrac{a_e}{3} & 1 - a_e & \dfrac{a_e}{3} \\ \dfrac{a_e}{3} & \dfrac{a_e}{3} & \dfrac{a_e}{3} & 1 - a_e \end{pmatrix},$$

Where a_e could vary for each edge e.

The Kimura 2-Parameter Model

Because of chemical similarities, the bases are classified as purines {A, G} and pyrimidines {C, T}. Assigning probability p_a to in-class changes (transitions), and p_b to out-of-class changes (transversions), we arrive at the Kimura 2-parameter model, with matrices

$$M^{(e)} = \begin{pmatrix} 1 - (a_e + 2b_e) & a_e & b_e & b_e \\ a_e & 1 - (a_e + 2b_e) & b_e & b_e \\ b_e & b_e & 1 - (a_e + 2b_e) & a_e \\ b_e & b_e & a_e & 1 - (a_e + 2b_e) \end{pmatrix},$$

where the rows and columns are ordered by the states A, G, C, T (purines, followed by pyrimidines). Typically $a > b$, since transitions are often observed more frequently than transversions.

The Kimura 3-Parameter Model

A slight generalization, introduced more for its mathematical structure than for biological reasons, is the Kimura 3-parameter model with transition matrices of the form

$$M^{(e)} = \begin{pmatrix} 1 - p_e & a_e & b_e & c_e \\ a_e & 1 - p_e & c_e & b_e \\ b_e & c_e & 1 - p_e & a_e \\ c_e & b_e & a_e & 1 - p_e \end{pmatrix},$$

Where $p_e = a_e + b_e + c_e$. A fundamental result on these structures is given by Hadamard conjugation [20] and [21] and it permits to introduce Fourier analysis as a tool for studying such models.

The Strand Symmetric Model

A strand symmetric Markov model is one whose mutation probabilities reflect the symmetry induced by the double-stranded structure of DNA [5]. In particular, a strand symmetric model for DNA must have the following equalities of probabilities in the root distribution:

$$\rho_A = \rho_T \quad \rho_C = \rho_G$$

And the transition matrices have the form

$$M^{(e)} = \begin{pmatrix} a_e & b_e & c_e & d_e \\ f_e & g_e & h_e & i_e \\ i_e & h_e & g_e & f_e \\ d_e & c_e & b_e & a_e \end{pmatrix},$$

That is, we have the following equalities of probabilities in the transition matrices

$$m_{AA} = m_{TT}, \quad m_{AC} = m_{TG}, \quad m_{AG} = m_{TC}, \quad m_{AT} = m_{TA},$$

$$m_{CA} = m_{GT}, \quad m_{CC} = m_{GG}, \quad m_{CG} = m_{GC}, \quad m_{CT} = m_{GA}.$$

Group-Based Models

Let T be a rooted tree with n taxa. We consider, on each vertex v, a random variable X_v which takes values on kl states where k is the cardinality of a finite abelian group G and l is a parameter of the model.

We can denote the states of the random variable by 2-tuples^j_i, where $j \in G$ and $i \in \{0, \ldots, l-1\}$. The entry $mi_{i_1,i_2}^{j_1,j_2}$ in a transition matrix represents the probability to pass from state $^{j_1}_{i_1}$ to state $^{j_2}_{i_2}$.

Definition 2.17: A phylogenetic model is a matrix-valued group-based model if, for each edge, the matrix transition probabilities satisfy

$$m^{j_1 j_2}_{i_1 i_2} = m^{k_1 k_2}_{i_1 i_2}$$

When $_{j1}-_{j2}=_{k1}-_{k2}$ (where the difference is taken in G) and the root distribution probabilities satisfy

$$\pi(r)_i^j = \pi(r)_i^k$$

For all $j, k \in G$.

Thus, the strand symmetric model of the previous section can be viewed as a group-based model. In fact, by the identification of the states

$$A = \begin{pmatrix} 0 \\ 0 \end{pmatrix} \quad G = \begin{pmatrix} 0 \\ 1 \end{pmatrix} \quad T = \begin{pmatrix} 1 \\ 0 \end{pmatrix} \quad C = \begin{pmatrix} 1 \\ 1 \end{pmatrix}.$$

The strand symmetric model becomes a matrix-valued group-based model with $l=2$ and $G=_{Z2}$.

Remark 2.18: Many probabilistic models of the mutation process – as evolution proceeds down a tree – focus on a single site in a sequence, and only on base substitutions occurring at that site. In general, we introduce more complicated models when we want to consider different types of sequence changes as insertions, deletions and inversions [24].

PHYLOGENETIC INVARIANTS

In 1987, Cavender and Felsenstein in [7], and, separately, Lake in [22] introduced the concept of phylogenetic invariants as a new approach to the study of phylogenetic trees arising from biological sequence data (i.e. the case of $k=4$ states: A, C, G, T). Obviously, we just consider the generalization to the case of k states, with k an arbitrary positive integer, $k \geq 2$.

We recall that, given a topological tree T with n leaves (or terminal taxa) and a model M of evolution along this tree, it is possible to compute the expected pattern frequencies of the k^n patterns of various states at the leaves, in terms of the parameters of the model, as explained in (2.2).

Definition 3.1: A phylogenetic invariant, for the topological tree T and the parameterized model M, is a polynomial in k^n variables, which becomes zero when the expected frequencies are substituted for the variables.

Since we want to consider an algebraic approach, we can work over the complex field. Thus, we will talk ofcomplex parameters to distinguish them from stochastic parameters, that is, positive real numbers. In both cases, we require that the root distribution and each row of the transition matrices sum to 1.

Let $\{z_{i1\cdots in}\}$ be a set of k^n indeterminates indexed by $1 \leq i_1, \ldots, i_n \leq k$, and denote by R the polynomial ring $C[z_{i1\cdots in}]$. We can restate the previous Definition in the following way.

Definition 3.2: A phylogenetic invariant, for the general Markov model (T, M), is a polynomial $f \in C[z_{i1\cdots in}]$ such that $f \equiv 0$ under the substitution $p_{i1\cdots in} \to z_{i1\cdots in}$ of the polynomial expressions for the expected pattern frequencies at the leaves.

Example 3.3: Since the $p_{i1\cdots in}$'s represent all the possible probabilities of the events in the joint distribution state, one has

$$\sum_{1 \leqslant i_1,\ldots,i_n \leqslant k} p_{i_1\cdots i_n} = 1.$$

$$(3.1)$$

This invariant, which is common to all n-taxon trees with k states, is called stochastic invariant.

Suppose that phylogenetic invariants can be found. This permits us to choose both the topological tree and the pararameterized model. In fact, starting from the observed data, we can compute the observed frequencies of patterns $\hat{p}_{i_1\cdots i_n}$'s. The observed data are distinct sequences of states (in particular, in the case k=4, these are aligned DNA sequences), one for each of the n species. All sequences have the same length M. Thus, the observed frequencies of patterns $\hat{p}_{i_1\cdots i_n}$'s are given by

$$\hat{P}_{i_1 \cdots i_n} = \frac{\text{occurrence of } i_1, \ldots, i_n}{M}.$$

Example 3.4: Consider four species with given DNA sequences of length 30.	
Species 1	CGTTACCCACTAGTTTATGACGTTACCCAC
Species 2	CGTTACCGACTAAATGCTGTCGTTACCGAC
Species 3	AGCCCCCCAATTATGAGCGTAGCCCCCCAA
Species 4	CGGGATTAAAATGCCGCGGGCGGGATTAAA

Thus, for example, one has $\hat{p}_{\text{TTCG}} = \dfrac{5}{30} = 0.16667$

If observed frequencies of patterns are good estimators of the expected frequencies, they will force the invariants to vanish or, at least, to be small. Thus, we choose the topological tree so that its invariants are close to vanish on the observed frequencies (then, in order to apply invariants to real data, one must decide what it means for an invariant to be "close to vanishing" on observed frequencies). More precisely, consider a phylogenetic invariant f   for a general Markov model (T, M),

where M is given in the unknown parameters $\pi(r)_i, m_{ij}^{(e)}$. Let $P = (_{pi1\cdots in})$ be the joint distribution tensor describing probabilities of states at the leaves. Hence f (P) =0. These probabilities are expressed in terms of the parameters of the modelM, that is, the entries $m_{ij}^{(e)}$ and the root distribution. Replacing P with a joint distribution tensor $_{P0}$, arising from any specific choice of parameters for T and M, one has again

$f(_{P0})=0$. Call $\hat{p}$ the tensor representing the observed pattern frequencies of the data. If T   and M are the correct tree and model relating to the sequence, then $\hat{p} \approx P_0$. Since $f(_{P0})=0$, then $f(\hat{p}) \approx 0$. Thus, the near vanishing of the phylogenetic invariants on the observed pattern frequencies is a good verification on T   and M as correct tree and model.

This model-based method of choosing topological trees will be useful if we are able, first of all, to produce "efficiently" phylogenetic invariants. Several techniques are used to find phylogenetic invariants:

the 4-point condition with log-det metric [7] and [27], the studying of algebraic relationships among expected frequencies [15] and [16], and harmonic analysis [13]. In general, in the cited works, these techniques are restricted to very specific topological trees and models (for example, in [7], the authors produce invariants for the Jukes–Cantor 2-base model with 4 terminal taxa) with few hopes to extend them to the general case. We suggest to look at the introduction of [1] for a better reference about these techniques.

Definition 3.5: Let (T_1, M_1) and (T_2, M_2) be two general Markov models with the same number of leaves and with, respectively, joint distribution tensors P_1 and P_2. If a polynomial f is such that $f(P_1) = 0$ (or $f(P_1) \approx 0$) but $f(P_2) \neq 0$ (or $f(P_2) \not\approx 0$) we say that f is topologically informative.

Example 3.6: Consider the model of Cavender and Felsenstein in [7]. This is a symmetric model with $k = 2$ states given by 0 and 1.

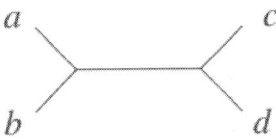

After the stochastic invariant, we have the linear invariants given by the symmetry of the model: $p_{0000} - p_{1111}$, $p_{0011} - p_{1100}$, $p_{0001} - p_{1110}$, $p_{0110} - p_{1001}$, etc. Finally, the last one is the informative invariant:

$$f = \left(p_{0100} + p_{1011} - p_{0111} - p_{1000}\right)\left(p_{0010} + p_{1101} - p_{0001} - p_{1110}\right) - \left(p_{0110} + p_{1001} - p_{0101} - p_{1010}\right)\left(p_{0000} + p_{1111} - p_{0011} - p_{1100}\right).$$

The origin of this invariant can be found in the 4-point condition for tree metrics. This polynomial vanishes only for the 4-leaf tree where a and b are neighbours, and does not vanish for generic joint distributions arising from the other two 4-leaf topologies (given by the other two ways to label leaves, with a in the same cherry of c or d). Thus f is topologically informative.

PHYLOGENETIC IDEAL AND PHYLOGENETIC VARIETY

As already expressed in Section 1, we will discuss here a thread of research in which the methods of Algebraic Geometry have been introduced to understand some of the probabilistic models used in Phylogenetics. The recent progress in understanding the set of possible probability distributions, arising from a model such as an algebraic variety, provides new theoretical results, and may point toward improved approaches to phylogenetic inference.

Since phylogenetic invariants are polynomials in the sthocastic parameters, we can consequently introduce the use of Algebraic Geometry. For this purpose, consider again a tree T with n leaves and kstates. First of all, recalling Definition 3.2, we observe that if two polynomials of $C[_{zi1\cdots in}]$ vanish under such substitution, then so do any of their linear combinations with coefficients in $C[_{zi1\cdots in}]$. From Chapter 0 of [10], it follows that the phylogenetic invariants form an ideal in R.

Definition 4.1: Let I_T be the ideal generated by the phylogenetic invariants of the general Markov model (T, M). I_T is the phylogenetic ideal of T.

We can introduce now a more geometric viewpoint. As already said in the previous section, a model M on a tree T, with n leaves, has $N := (k-1)+(2n-3)k(k-1)$ free parameters. Thus, the stochastic parameter space for the tree T is given by $S \subset [0,1]^N$, and each $s \in S$ represents a model $M = (\pi(r), \{M^{(e)}\})$. Using Formula (2.2), we can define a parameterization map

$$\phi_T : S \to [0,1]^{kn}, s \to P = [_{p11}\cdots_1, \ldots,_{pkk}\cdots_k], \tag{4.1}$$

where $[0,1]^{kn}$ represents the joint distribution state. An element $P \in [0,1]^{kn}$ in the image of ϕ_T represents a joint distribution of pattern frequencies at the leaves of T. Since, by Formula (2.2), ϕ_T is a polynomial map in the unknown parameters, we can extend it to

$$\Phi_T : \mathbb{C}^N \to \mathbb{C}^{kn}. \tag{4.2}$$

Definition 4.2: Let V_T be the (Zarisky) closure of the image of Φ_T, that is $V_T = \overline{\Phi T(\mathbb{C}^N)}$. V_T is called the phylogenetic variety associated to the tree T.

Roughly speaking, V_T is a variety that contains the (complex) joint distribution for all possible choices of (complex) numerical parameters $M = (\pi(r), \{M^{(e)}\})$ of general Markov model on the tree T. In applications, the tree topology is usually the parameter of greatest interest. If an observed distribution of pattern frequencies were "close" to V_T, that could be interpreted as support for inferring T. Thus, phylogenetic invariants allow the inference of T without having to estimate all the other parameters, as, on the contrary, maximum likelihood requires.

Remark 4.3: Extending the parameterization Φ_T to the complex numbers from the stochastic setting is done because an algebraically closed field provides the easiest and most natural setting for understanding polynomial maps. Of course, complex parameters and complex joint distributions are not so natural from a biological or statistical viewpoint. Obviously, the final goal would be to understand the model in the stochastic setting.

We have the following method which is statistically consistent.

INPUT: the joint distribution of observed frequencies $\hat{p}$.

i.   fix M;
ii. for each tree T

• Find some/most/all invariants f for V_T;

• Test if $f(\hat{p}) = 0$.

OUTPUT: the tree T for which $\hat{p}$ is as close as possible to V_T.

Understanding V_T well means both theoretical and practical understanding of problems of phylogenetic inference. One part of understanding V_T is describing it implicitly, as the zero set of polynomials. This means to find polynomials $f \in R$ such that $f(q)=0$ for all $q \in V_T$. This is equivalent to find the kernel of the map

$$\tilde{\Phi}_T : \mathbb{C}[z_{i_1 \cdots i_n}] \to \mathbb{C}[s_1, \ldots, s_N],$$

where $\mathbb{C}[s_1, \ldots, s_N]$ is the polynomial ring associated to vector space $\mathbb{C}^N$ (we can take, as variables s_i's, for example, the unknown stochastic parameters of the Markov model, $\pi(r)_i$, $m_{ij}^{(e)}$). The kernel of $\tilde{\Phi}_T$ is the ideal I of the polynomials in the $z_{i1 \cdots in}$'s vanishing for all choices of (stochastic or complex) parameters s_i's, i.e. it corresponds exactly to the phylogenetic ideal as defined in Definition 4.1.

What we want to do is to find explicitly the ideal of a general Markov model for each topological tree. Since, by Hilbert Basis Theorem [10, Theorem 1.2, p. 27], these ideals are finitely generated, this is equivalent to giving a list of generators of each ideal. The research of phylogenetic invariants seems to be a specific issue of computational Algebraic Geometry. Here, the main techniques to manipulate polynomials are given by Gröbner Basis [10, Chapter 15]. Theoretically, Gröbner basis permit to find all the phylogenetic invariants of a topological tree with a parameterized model. Unluckily, in the practical situation the use of Gröbner Basis is limited to a small number of leaves and states. In fact, the basic algorithm to give I_T is the application of the elimination process to the set of equations $z_{i1 \cdots in} - p_{i1 \cdots in} = 0$ with respect to the indeterminates/parameters $\pi(r)_i$'s, $m_{ij}^{(e)}$'s. This process involves k^n polynomials of degree $2n-3$ in the variables $\pi(r)_i$'s, $m_{ij}^{(e)}$'s. Thus, as soon as the number of leaves, or the number of states grows, the computation becomes more and more complex and technology, at the moment, is not able to produce any results.

Remark 4.4: In finding phylogenetic invariants, the main goal is to determine the full ideal I_T, i.e. an ideal-theoretic definition of V_T. However, a weaker goal is to determine a set of polynomials whose zero set is V_T. Namely, what we are looking for is a set-theoretic definition of the variety without determining the whole ideal.

Researchers, in the field of Phylogenetic Algebraic Geometry, are looking for different techniques and tools to find phylogenetic invariants.

In [1], Allman and Rhodes present several methods of finding phyloge-netic invariants for the general Markov model of base substitution along any topological tree. In particular, the authors do not require any con-ditions on the numbers n and k of leaves and states. The constructions are based on commutation and symmetry relations of matrix expres-sions and that requires only linear algebra. In particular, for a 3-taxon tree T, a strong set of invariants can be found. These invariants have degree $k+1$ (the lowest possible) and have many terms. For example, if $k=4$, they are all degree five invariants (1728-dimensional space) and the number of terms in each invariant is about 180. The construc-tions are successively generalized to the n-taxon case. Unluckily, here, these invariants do not generate the full ideal, but in some cases they give a satisfactory result. It is important to observe that, since they are expressed in matrix form, invariants may be evaluated through numeri-cal linear algebra.

A particular interesting issue concerns group-based models on an ar-bitrary tree, for which Sturmfels and Sullivant have given a complete description of the phylogenetic ideal. Their proof is based on the Ha-damard conjugation which consists in a change of variables making the parameterization ϕ_T simpler. In this case, such parameterization is given by monomials. Varieties parameterized by mononials are called toric varieties and are a well-understood class in Algebraic Geometry. The main results are:

Theorem 4.5 Sturmfels and Sullivant [28]
For any group-based model on a phylogenetic tree T, the prime ideal of phylogenetic invariants is generated by the invariants of the local submodels around each interior node of T, together with the quadrics which encode conditional independence statements along the splits of T.

Theorem 4.6 Sturmfels and Sullivant [28]
Let T be an arbitrary binary rooted tree. Modulo the stochastic invari-ant,

a. the ideal of the Jukes – Cantor binary model is generated by polynomials of degree 2;

b. the ideal of the Jukes – Cantor DNA model is generated by polynomials of degree 1, 2 and 3;

c. the ideal of the Kimura 2-parameter model is generated by polynomials of degree 1, 2, 3 and 4;

d. the ideal of the Kimura 3-parameter model is generated by polynomials of degree 2, 3 and 4.

Each of these generating sets has an explicit combinatorial description and it is a Gröbner basis.

A fully observable homogeneous Markov model has no hidden nodes and all matrices are the same. An explicit analysis of this kind of model, on a tree with at most five leaves and $k=2$, can be found in [11].

Remark 4.7: Although we consider here only the part of the study of phylogenetic invariants concerned with Algebraic Geometry, it is mandatory to recall that a statistical understanding of the behaviour of these polynomials is necessary. This is due, in particular, to the fact that we want to apply invariants to real and noisy data. Moreover, the models of evolution are only an approximation of reality, then, from a statistical point of view, we need the robustness of method under violation of model assumptions. Finally, Statistics will be necessary for a good defintion of "close to vanish".

We now fix our attention on the technique of flattening of a n-dimensional tensor. As we will see in the next two Sections, the flattenings permit to obtain invariants from the "local" structure of the tree. Moreover, the flattenings are strictly connected with the theory of secant varieties. There is a huge literature on this theory; some papers are becoming fundamental in the research of phylogenetic invariants.

FLATTENINGS

We introduce, now, the notion of edge flattening of a tensor $P \in C^{kn}$ according to an n-taxon tree T. Let $P = \Phi_T(s)$ be the joint distribution of states for some parameters choice for the general Markov model on T. Consider an edge e of T. Then e induces a split of the taxa according to the connected components of $T \geq \{e\}$. We can assume, eventually re-ordering the indices in P, that the split is $\{\{i_1, \ldots, i_t\}, \{i_{t+1}, \ldots, i_n\}\}$. We can imagine now a statistical model based on the split induced by e : we can group the taxa $\{i_1, \ldots, i_t\}$ and the taxa $\{i_{t+1}, \ldots, i_n\}$, so that each is on a leaf attached to a common ancestral node, which is choosen to be on one vertex of e (Figure.6). The joint states at the taxa $\{i_1, \ldots, i_t\}$, are described through a single k^t-state variable and similarly at $\{i_{t+1}, \ldots, i_n\}$ are described through a single k^{n-t}-state variable. Thus, we have a coarser graphical model with one hidden k-state internal node and two descendant nodes with, respectively, k^t and k^{n-t} states. Forming the joint distribution for this coarser model, we get a $k^t \times k^{n-t}$ matrix $_{Flate}(P)$ which is defined in the following way: fix any ordering of $_{j1} := [k]^t$ and $_{j2} := [k]^{n-t}$ and for $u \in _{j1}$, $v \in _{j2}$ let

$$_{Flate}(P)(u,v) = P(_{u1}, \ldots, _{ut}, _{v1}, \ldots, _{vn-t}).$$

Then $_{Flate}(P)$ can be seen as a joint distribution for a related graphical model with a less complicated structure: one hidden k-state internal node and two descendant nodes with k^t and k^{n-t} states, respectively.

Example 5.1: Consider the following 5-leaf tree T

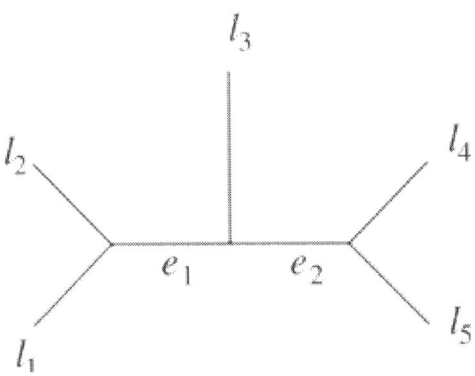

With $k=2$ states at each vertex, represented by 0 and 1. The two splits

$$\{\{l_1, l_2\}, \{l_3, l_4, l_5\}\} \quad \text{and} \quad \{\{l_1, l_2, l_3\}, \{l_4, l_5\}\}$$

Give, respectively, the flattenings

$$\text{Flat}_{e_1}(P) = \begin{pmatrix} p_{00000} & p_{00001} & p_{00010} & p_{00011} & p_{00100} & p_{00101} & p_{00110} & p_{00111} \\ p_{01000} & p_{01001} & p_{01010} & p_{01011} & p_{01100} & p_{01101} & p_{01110} & p_{01111} \\ p_{10000} & p_{10001} & p_{10010} & p_{10011} & p_{10100} & p_{10101} & p_{10110} & p_{10111} \\ p_{11000} & p_{11001} & p_{11010} & p_{11011} & p_{11100} & p_{11101} & p_{11110} & p_{11111} \end{pmatrix} \tag{5.1}$$

And

$$\text{Flat}_{e_2}(P) = \begin{pmatrix} p_{00000} & p_{00001} & p_{00010} & p_{00011} \\ p_{00100} & p_{00101} & p_{00110} & p_{00111} \\ p_{01000} & p_{01001} & p_{01010} & p_{01011} \\ p_{01100} & p_{01101} & p_{01110} & p_{01111} \\ p_{10000} & p_{10001} & p_{10010} & p_{10011} \\ p_{10100} & p_{10101} & p_{10110} & p_{10111} \\ p_{11000} & p_{11001} & p_{11010} & p_{11011} \\ p_{11100} & p_{11101} & p_{11110} & p_{11111} \end{pmatrix}. \tag{5.2}$$

Here, for example, the $(01,000)$-entry of $\text{Flat}_{e_1}(P)$ is the probability of observing state 01 at leaf$\{l_1, l_2\}$, and state 000 at leaf $\{l_3, l_4, l_5\}$. Of course, this entry is precisely p_{01000}.

Remark 5.2: A combinatorial result, the Splits Equivalence Theorem, states that a tree is uniquely determined by its set of splits. See [28] for a proof.

For the coarser graphical model, the joint distribution matrix must have the form

$$\text{Flat}_e(P) = M_1^T \text{diag}(\pi(r)) M_2 \tag{5.3}$$

Where M_1 and M_2 are $k \times k^t$ and $k \times k^{n-t}$ matrices and $\text{diag}(\pi(r))$ is a diagonal matrix with (i,i)-entry $\pi(r)_i$ (the precise description of M_1 and M_2 can be found in [4]).

Remark 5.3: The coarser model is not a phylogenetic tree since the number of states at the two leaves are different powers of k.

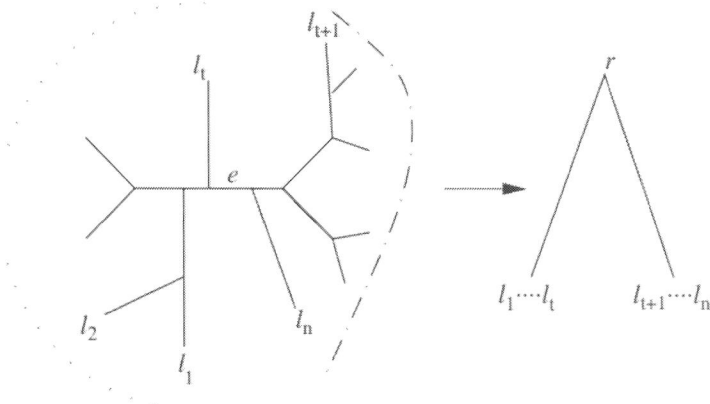

Figure.6: An example of edge flattening.

From (5.3) we quickly obtain that $\text{rank}(_{\text{Flate}}(P)) \le k$. Hence, all $(k+1) \times (k+1)$ minors of $_{\text{Flate}}(P)$ must vanish. These minors generate the full ideal of polynomials vanishing on matrices of rank $\le k$, and thus we have found all phylogenetic invariants associated to the coarser model. Such invariants are also invariants for the original model on T, and they are called edge invariants associated to the edge e. Moreover, we denote by $_{\text{Fedge}}(T)$ the set of all $(k+1) \times (k+1)$ minors of the edge flattenings$_{\text{Flate}}(P)$ as e varies on E. An important result concerning flattenings is the following

Theorem 5.4 Allman and Rhodes [3]
For $k=2$ and any number of taxa n, the phylogenetic ideal $_{\text{IT}}$, for the general Markov model M on an n-taxon tree T, is generated by $_{\text{Fedge}}(T)$, the 3×3 minors of all edge flattenings of a $2 \times 2 \times \cdots \times 2$ tensor of indeterminants.

Thus, in particular one has

Corollary 5.5: For the 5-leaf tree T of Example 5.1, $_{\text{IT}}$ is generated by all 3×3 minors of matrices (5.1) and (5.2).

However, for a larger k , it is not enough to consider only 2-dimensional edge flattenings (i.e., flattenings to matrices) to obtain generators of the phylogenetic ideal. Consider, for example, a 3-taxon tree T:

if $k > 2$ we know that $_{IT}$ contains polynomials of degree $k+1$ although $_{Fedge}(T)$ is empty. Hovewer, we can give an interesting result.

Proposition 5.6 Allman and Rhodes [3]
For any k-state general Markov model on T, or any submodel, the phylogenetic ideal contains all edge invariants.

To find other invariants, we can introduce another kind of flattening which produces 3-dimensional tensors. Consider an internal node v of T, contained in edges $_{e1'e2'e3}$. Then v induces a tripartition of the taxa according to the connected components of $T \setminus \{v,_{e1'e2'e3}\}$. We may assume the tripartition is

$$\{\{_{|1'}\cdots,_{|n1}\},\{_{|n1+1'}\cdots,_{|n1+n2}\},\{_{|n1+n2+1'}\cdots,_{|n1+n2+n3}\}\},$$

Where $_{n1}+_{n2}+_{n3}=n$. Then a vertex flattening of P at v is a $(k_{n1} \times k_{n2} \times k_{n3})$-array $_{Flatv}(P)$ defined as follows: fix an ordering of $_{J1}=[k]_{n1'}$ $_{J2}=[k]_{n2'}$ $_{J3}=[k]_{n3'}$ and for $x \in_{J1}$, $y \in_{J2}$, $z \in_{J3}$ let

$$_{Flatv}(x,y,z)=P(_{x1'}\cdots,_{xn1'y1'}\cdots,_{yn2'z1'}\cdots,_{zn3}).$$

Thus, the final result of a vertex flattening is a graphical model with one hidden k-state internal node and three descendant nodes with k_{n1}, $k_{n2'}$ and k_{n3} states, respectively. Since an ideal is associated to such a graphical model, we can talk of the ideal of the vertex flattening (Figure.7).

We have now all the elements to give the following

Conjecture 5.7 Allman and Rhodes [3].
For any k and any number of taxa n, the phylogenetic ideal $_{IT'}$ for the general Markov model on an n-taxon tree T, is the sum of the ideals associated to the flattenings of P at vertices of T.

It is important to remark that this Conjecture, for $k=2$, is identical to Theorem 5.4. In fact, by the results of Lansdberg and Manivel [23] we know that, in this case, the ideal associated to a vertex flattening is the sum of the ideals associated to the edge flattenings of the three edges containing the vertex.

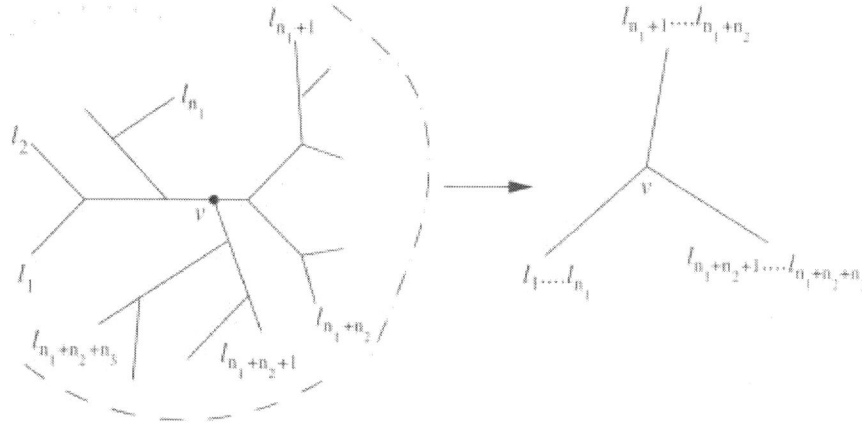

Figure.7: An example of vertex flattening.

SECANT VARIETIES

In the previous section we only considered matrices with rows that sum to 1. This probabilistic condition can be interpreted in Algebraic Geometry as the fact that each row of a transition matrix is an element of a certain affine subspace of a projective space P^{k-1}. At the same time, we can view $_{VT}$ projectively. In fact, by the stochastic invariant, one has $_{VT} \subset P^{kn-1}$. The passage to the projective case forces us to look only for phylogenetic invariants among homogeneous polynomials.

Consider a 3-taxon rooted star tree T in the projective setting for k states. Let r be the root of T and let $_{e_i}$ the edge connecting r to $_{l_i}$, $i=1$, 2, 3 (Figure.8). Suppose that the state at r is momentarily fixed as $\tilde{k}$. Then, for each edge leading away from r, towards a leaf, we have a point $\overline{m}_{\tilde{k}l_i} = (m_{\tilde{k}1}^{(e_i)}), \cdots, m_{\tilde{k}k}^{(e_i)} \in p^{k-1}$ that represents the $\tilde{k}$ th row of the transition matrix $M^{(e_i)}$. Thus, if we define

$$P^{\tilde{k}} := \overline{m}_{\tilde{k}l_1} \otimes \overline{m}_{\tilde{k}l_2} \otimes \overline{m}_{\tilde{k}l_3} \in P^{k-1} \times P^{k-1} \times P^{k-1},$$

then $P^{\hat{k}}$ is a point in the Segre product [18, Example 2.11, p. 25] of three projective spaces whose entries (up to scaling) are the expected frequencies of observing pattern conditioned by the state at the root being $\tilde{k}$. Summing over all possible states at r, we obtain the joint distribution

$$P = P^1 + P^2 + \cdots + P^k.$$

Since we are summing k points on the variety $P^{k-1} \times P^{k-1} \times P^{k-1}$, we obtain $P \in _{VT} = Sec^{k-1}(P^{k-1} \times P^{k-1} \times P^{k-1})$, i.e. the (k-1)-secant variety [18, Example 8.5, p. 90] of the Segre product of three P^{k-1}.

We have to point out that the root distribution does not explicitly appear, since it has been accounted for in the arbitrary scaling factors that appear in each P^i, when we choose particular projective coordinates to express them. Hence, the joint distribution P has been decomposed as the sum of k rank 1 tensors, one for each possible state at the root (and this forces P to have rank k, by the definition itself of tensor rank).

Example 6.1: Consider $k=2$. The general Markov model on $_{T3}$ has only 7 parameters and since the stochastic invariant cuts out a 7-dimensional subspace of C^{2^3}, one could expect that there are no other invariants. In fact, one has $Sec^1(P^1 \times P^1 \times P^1) = P^7$ (i.e. every $2 \times 2 \times 2$ tensor is in the closure of the rank 2 tensors). We want to underline that for the

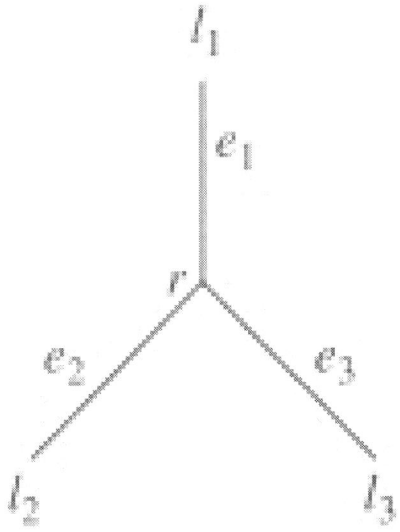

Figure.8: The 3-taxon tree $_{T3}$.

3-taxon tree, the construction of edge invariants yields nothing, since there are no internal edges.

Example 6.2: Go on considering $_{T3}$, but with $k=3$. In this case the ideal defining $Sec^2(P^2 \times P^2 \times P^2)$ was found in [17], and given in terms of Bayes models. Let $A = (_{pij1})$, $B = (_{pij2})$ and $C = (_{pij3})$ be three 3×3-matrices obtained by taking slices of the 3-dimensional tensor P associated to the model. Then one has

Proposition 6.3 Garcia et al [17]: Let I be the ideal of $Sec^{k-1}(P^2 \times P^2 \times P^2)$, the naive Bayes model with $n=3$ ternary features with k classes. If $k=2$ then I is generated by the 3×3-subdeterminantal of any two-dimensional table obtained by flattening the 3-dimensional tensor P. If $k=3$ then I is generated by the quartic entries of the various 3×3-matrices of the form

$A \cdot adj(B) \cdot C - C \cdot adj(B) \cdot A.$

If $k=4$ then I is the principal ideal generated by the following homogeneous polynomial of degree 9 with 9,216 terms: $det(B)^2 \cdot det(A \cdot B^{-1} \cdot C - C \cdot B^{-1} \cdot B)$.

If $k=5$ then I is the zero ideal.

More generally, let T be a star tree with an internal node r and n leaves $_{li}$. We can suppose that the hidden variable associated to r has k states, while the hidden variable at the leaf $_{li}$ has $_{\alpha i}$ states, $i=1,\ldots,n$. The variety associated to this model is the $(k-1)$-secant variety of the Segre product $P_{\alpha1}^{-1} \times P_{\alpha2}^{-1} \times \cdots \times P_{\alpha n}^{-1}$. From now on we will use the notation $V(k; _{\alpha1,\alpha2,\ldots,\alpha n})$ to denote $Sec^{k-1}(P_{\alpha1}^{-1} \times P_{\alpha2}^{-1} \times \cdots \times P_{\alpha n}^{-1})$.

There is a useful relationship between the varieties $V(k; _{\alpha1,\alpha2,\ldots,\alpha n})$ and $V(k;k,k,\ldots,k)$. In fact, for any joint distribution $P \in V(k;k,k,\ldots,k)$, there is an "action" by $k \times _{\alpha n}$ complex matrices M in the last index of P. This gives us a point $P *_n M \in V(k;k,k,\ldots,k,_{\alpha n})$ (here k is repeated n-1 times). Using the map $_{\Phi T}$, if $P = _{\Phi T}(\pi(r), \{_{M1,M2,\ldots,Mn}\})$, where $_{Mn}$ is the matrix on the edge leading to the n-th leaf, then $P *_n M = _{\Phi T}(\pi(r), \{_{M1,M2,\ldots,Mn-1,Mn}M\})$, though the action extends to the points (on the variety) that are not in the image of $_{\Phi T}$. We can define also an "action" by $_{\alpha n}\times k$ matrices N on $V(k;k,k,\ldots,k,_{\alpha n})$. Thus, given a $k\times_{\alpha n}$ matrix M and an $_{\alpha n}\times k$ matrix N we have maps

$$V(k;k,k,\ldots,k) \underset{*_n N}{\overset{*_n M}{\underset{\longleftarrow}{\longrightarrow}}} V(k;k,k,\ldots,k,\alpha).$$

(6.1)

From these maps, we can obtain maps between the ideals of the two varieties. The compositions of these maps are related to $GL(k,C)$ and $GL(_{\alpha n},C)$ actions. In a similar way, we can define an action on each distinct index, not just on the last one. Thus, we obtain an action of $GL(_{\alpha 1},C)\times GL(_{\alpha 2},C)\times\cdots\times GL(_{\alpha n},C)$ on $V(k;_{\alpha 1'\alpha 2'}\ldots,_{\alpha n})$. We have the following

Theorem 6.4 Allman and Rhodes [3]
Suppose $_{\alpha 1'\alpha 2'}\ldots,_{\alpha n}\geq k$. Let F be any set of polynomials whose zero set is $V(k;k,k,\ldots,k)$. For $t=1,2,\ldots,n$, let $Z_t=(z_{ij}^t)$ be $_{\alpha t}\times k$ matrices of indeterminants. For an $(_{\alpha 1}\times_{\alpha 2}\times\cdots\times_{\alpha n})$-tensor P of indeterminants, let $\tilde{P}$ be the $(k\times k\times\cdots\times k)$-tensor that results from letting each $_{zt}$ acts formally in the tth index of P (i.e. as the lower map in (6.1)). Let $\hat{F}$ denote the set of polynomials in the entries of P obtained from those in F by substituting into them the entries of P, expressing the results as polynomials in the z_{ij}^t, and then extracting the coefficients. Let $_{Fedge}$ be the set of $(k+1)\times(k+1)$ minors of the n flattenings of P on edges of the star tree. Finally, let $\hat{F}(k;x_1,x_2,\ldots\ldots x_n)=\hat{F}\cup\hat{F}edge$. Then $F(k;_{\alpha 1'\alpha 2'}\ldots,_{\alpha n})$ defines $V(k;_{\alpha 1'\alpha 2'}\ldots,_{\alpha n})$ set-theoretically.

Similarly, an ideal-theoretic version of this result can be given:

Theorem 6.5 Allman and Rhodes [3]
Suppose $_{\alpha 1'\alpha 2'}\ldots,_{\alpha n}\geq k$. Let F be any set of polynomials generating the ideal of $V(k;k,k,\ldots,k)$. Then the set $F(k;_{\alpha 1'\alpha 2'}\ldots,_{\alpha n})$ generates the ideal of $V(k;_{\alpha 1'\alpha 2'}\ldots,_{\alpha n})$.

Remark 6.6: Although Allman and Rhodes define $V(k;_{\alpha 1'\alpha 2'}\ldots,_{\alpha n})$ as the variety associated to a star tree with t leaves, we obviously have to consider only $n=3$, because the tree is bifurcating. We have to point out that a set of polynomials, defining the variety $V(k; \alpha_1, \alpha_2, \ldots, \alpha_n)$, when $k=2$, can be found in [23].

Consider a vertex flattening on a tree T. Now, the variety associated to the coarsened model is $V(k; k^{n_1}, k^{n_2}, k^{n_3})$, i.e. the variety of rank k tensors of size $k^{n_1} \times k^{n_2} \times k^{n_3}$. It is important to observe that, in such a case, one has $\alpha_i = k^{n_i}$ for an integer n_i depending on the splitting. Thus, the hypotheses $\alpha_i \geq k$ are satisfied and, in some way, we can try to use polynomials vanishing on $V(k; k, k, k)$ to obtain invariants for the vertex flattening $V(k; k^{n_1}, k^{n_2}, k^{n_3})$ and then for the starting tree T. More precisely, one has the following

Theorem 6.7 Allman and Rhodes [3]
For a 3-leaf star tree, let F be a set of polynomials defining $V(k; k, k, k)$ set-theoretically, and let $F(k; \alpha_1, \alpha_2, \alpha_3)$ be as defined in Theorem 6.4. For an n-taxon tree T, let $F(T)$ be the union of all sets $F(k; k^{n_1}, k^{n_2}, k^{n_3})$ associated to 3-dimensional flattenings at nodes of T. Then the zero set of $F(T)$ is the phylogenetic variety V_T.

In several cases, edge flattenings and vertex flattenings permit to determine, at least set-theoretically, the phylogenetic variety. We can then investigate if different kinds of flattenings can give new phylogenetic invariants (see [3, p. 12]). In any case, the previous theorems seem to suggest that the phylogenetic variety is determined by the local structure of the tree and encourages Phylogenetic Algebraic Geometry in this direction.

We can conclude with a remark about Theorem 6.4. An important consequence of this Theorem is the following

Corollary 6.8: For $n \leq 5$, the ideal of the variety $V(2; \alpha_1, \alpha_2, \ldots, \alpha_n)$ associated to the hidden naive Bayes model with a 2-state hidden variable and n observed variables with $\alpha_1, \ldots, \alpha_n$ states, is generated by the 3×3 minors of all 2-dimensional flattenings associated to bipartitions of the observed variables.

We have to point out that the case $n=3$ was already proved in [23]. The previous Corollary solves several cases of the following

Conjecture 6.9 Garcia et al. [17]

The prime ideal I_{QG} of any naive Bayes model G with $r=2$ classes is generated by the 3×3-subdeterminants of any 2-dimensional table obtained by flattening the n-dimensional table $(p_{i1 i2 \cdots in})$.

Theorem 6.4 limits the study of the ideal of $V(2; \alpha_1, \alpha_2, \ldots, \alpha_n)$ to the "simplest" case $V(2; 2, 2, \ldots, 2)$, since, applying the construction of this Theorem and maps as in (6.1), we obtain the generators for $V(2; \alpha_1, \alpha_2, \ldots, \alpha_n)$, with $\alpha_i \geq 2$. Thus, the Conjecture can be restated as

Conjecture 6.10 Garcia et al. [17]

The ideal of the variety $V(2; 2, 2, \ldots, 2)$, that is, the ideal associated to the hidden naive Bayes model with a 2-state variable and n 2-state observed variables, is generated by the 3×3-subdeterminants of all two-dimensional flattenings arising from bipartitions of the observed variables.

ACKNOWLEDGMENTS

The author would like to thank A. Alzati, M. Bertolini, G. Casnati, C. Fontanari, A. Gimigliano, R. Notari, G. Ottaviani and P. Freguglia, for the opportunity they gave him to talk about Phylogenetic Algebraic Geometry in their Departments. Special thanks go to Prof. Kristian Ranestad who introduced the author to this interesting and beautiful research area and to Pier Angelo Lenzi who suggested several corrections to this paper.

This research was partially supported by GNSAGA of INdAM (Italy).

REFERENCES

1. E.S. Allman, J.A. Rhodes, Phylogenetic invariants for the general Markov model of sequence mutation, Math. Biosci. 186 (2003) 113–144.
2. E.S. Allman, J.A. Rhodes, Phylogenetic invariants for stationary base composition, J. Symbolic Comput., 41 (2) (2006) 138–150.
3. E.S. Allman, J.A. Rhodes, Phylogenetic ideals and varieties for the general Markov model, Adv. in Appl. Math., (2007), to appear (arXiv:math.AG/0410604).

4. E.S. Allman, J.A. Rhodes, Phylogenetics, Modeling and Simulation of Biological Networks, Proc. Sympos. Appl. Math., Amer. Math. Soc., Providence, RI (2007), to appear.

5. M. Casanellas, S. Sullivant, The strand symmetric model, in: L. Pachter, B. Sturmfels (Eds.), Algebraic Statistics for Computational Biology, Cambridge University Press, Cambridge, 2005, pp. 305–321.

6. M.V. Catalisano, A.V. Geramita, A. Gimigliano, Ranks of tensors, secant varieties of Segre varieties and fat points, Linear Algebra Appl. 355 (2002) 263–285.

7. J.A. Cavender, J. Felsenstein, Invariants of phylogenies in a simple case with discrete states, J. Class 4 (1987) 57–71.

8. D. Cox, J. Little, D. O'Shea, Ideals, Varieties, and Algorithms, Undergraduate Texts in Mathematics, Springer, New York, 1997.

9. D. Cox, J. Little, D. O'Shea, Using Algebraic Geometry, Springer, New York, 1998.

10. D. Eisenbud, Commutative Algebra with a View Toward Algebraic Geometry, Graduate Texts in Mathematics, Springer, New York, 1994.

11. N. Eriksson, Toric ideal of homogeneous phylogenetic models, in: ISSAC 2004, ACM, New York, 2004, pp. 149–154.

12. N. Eriksson, K. Ranestad, B. Sturmfels, S. Sullivant, Phylogenetic algebraic geometry, in: C. Ciliberto, A.V. Geramita, B. Harbourne, R.M. Miro-Roig, K. Ranestad (Eds.), Projective Varieties with Unexpected Properties: A volume in Memory of Giuseppe Veronese; Proceedings of the international Conference 'Varieties with Unexpected Properties', Siena, Italy, June 8–13, 2004, Walter de Gruyter, 2005, pp. 177–197.

13. S.N. Evans, T.P. Speed, Invariants of some probability models used in phylogenetic inference, Ann. Statist. 21 (1) (1993) 355–377.

14. J. Felsentein, Inferring phylogenies, Sinauer Associates, Inc., Sunderland, 2003.

15. V. Ferretti, D. Sankoff, The empirical discovery of phylogenetic invariants, Adv. Appl. Probab. 25 (2) (1993) 290–302.

16. V. Ferretti, D. Sankoff, Phylogenetic invariants for more general evolutionary models, J. Theor. Biol. 173 (1995) 147–162.

17. L.D. Garcia, M. Stillman, B. Sturmfels, Algebraic geometry of Bayesian network, J. Symbolic. Comput. 39 (2005) 331–355.

18. J. Harris, Algebraic Geometry: A First course, Graduate Texts in Mathematics, vol. 133, Springer, Berlin, 1992.

19. R. Harthshorne, Algebraic Geometry, Graduate Texts in Mathematics, Springer, New York, 1977.

20. M.D. Hendy, The relationship between simple evolutionary tree models and observable sequence data, Syst. Zool. 38 (1989) 310–321.

21. M.D. Hendy, D. Penny, A framework for the quantitative study of evolutionary trees, Syst. Zool. 38 (1989) 297–309.

22. J.A. Lake, A rate independent technique for analysis of nucleic acid sequence: evolutionary parsimony, Mol. Biol. Evol. 4 (1987) 167–191.

23. J.M. Landsberg, L. Manivel, On the ideal of secant varieties of Segre varieties, Found. Comput. Math. 4 (4) (2004) 397–422.

24. L. Pachter, B. Sturmfels, The Mathematics of Phylogenomics, preprint, 2005.
25. L. Pachter, B. Sturmfels (Eds.), Algebraic Statistics for Computational Biology, Cambridge University Press, Cambridge, 2005.
26. C. Semple, M. Steel, Phylogenetics, Oxford Lecture Series in Mathematics and its Applications, vol. 24, Oxford University Press, Oxford, 2003.
27. M. Steel, Recovering a tree from the leaf colourations it generates under a Markov model, Appl. Math. Lett. 7 (2) (1994) 19–24.
28. B. Sturmfels, S. Sullivant, Toric ideals of phylogenetic invariants, J. Comput. Biol. 12 (2) (2005) 204–228.

CITATION

Cristiano Bocci, Dipartimento di Matematica, Università di Milano, Via Cesare Saldini 50, 20133 Milano, italy, topics on phylogenetic algebraic geometry, doi:10.1016/j.exmath.2007.02.001.

Lefschetz thus served to move the American algebraic school in the same direction Scorza and Rosati had gone in Italy, namely, toward an exploration of endomorphisms of Abelian varieties, i.e., multiplication of Riemann matrices, in a purely algebraic, as opposed to topological, way. Although Albert moved from pure algebra to algebraic geometry and Scorza moved from algebraic geometry to pure algebra, the two approaches they represented coalesced for a decade. Albert made it quite clear: "The structure of impure Riemann matrices was shown by G. Scorza to be expressible in terms of the structure of its pure submatrices, and the case where D [the multiplication algebra of a Riemann matrix] is a commutative algebra was discussed at length by S. Lefschetz. No further really important work in this direction was made until that of a recent paper by C. Rosati. … We give a rigorous algebraic proof of Rosati's main results and obtain his theorems, not in terms of unknown sub-algebras of given algebras B but actually in terms of the central field of B and its rank function" [Albert, 1931, 1]. Albert's work linked up with Scorza's during the 1930s. By that time, however, the Italian school of algebraic geometry was no longer capable of actively following the new developments spurred by Albert's works.

OSCAR ZARISKI: FROM UNEASINESS TO CRITICISM

When the 22-year-old Oscar Zariski came from Russia to Rome in 1921, Lefschetz was renewing his Italian connections.[10] Zariski remained in Rome until 1927, studying algebraic geometry mostly under Castelnuovo's supervision and learning both scientific and life lessons from his long and frequent conversations with Federigo Enriques. Zariski may thus be considered a student of both Castelnuovo and Enriques. His mathematical "style" and the methods he used differed little from those of the most brilliant of his Italian mathematical contemporaries—Luigi Campedelli, Oscar Chisini, Fabio Conforto, and many others.

By the close of the 1920s, however, a major shift occurred. While Chisini, Campedelli, and later Beniamino Segre and others did not feel the need—and were not encouraged by their teachers—to renovate the methods and ideas inherited from their masters, Zariski felt increasingly uneasy working in that tradition. In 1928, still considering himself a close adherent to Castelnuovo's methods, he attended the International Congress of Mathematicians held in Bologna. Writing to his wife, he gloried in his association with his mentor's mathematical thought: "Today Castelnuovo gave a lecture: it was the best so far and it made a great impression on everyone, both for its content and for its elegant style. It was a real work of art. He did me the greatest honour of interrupting his lecture at a certain point ... in order to announce to the audience my upcoming communication to the Congress, in which, according to him, I have made an important step toward the solution of a fundamental problem which is still unresolved ... you can understand how significant a sign of recognition it was" [Parikh, 1991, 60].

But 1928 also witnessed the first symptoms of Zariski's uneasiness with the Italian approach to algebraic geometry. It was apparently Castelnuovo himself who sparked this unease. He reputedly declared that "[t]he methods of the Italian school have reached a dead end and are inadequate for further progress in the field of algebraic geometry" [Parikh, 1991, 36]. If Castelnuovo did, indeed, make this pronouncement, it is easy to understand why he suggested that his young student should go further in his studies and "explore the work of Solomon Lefschetz" [Parikh, 1991, 36]. In the second half of the 1920s, Castelnuovo thus viewed the American approach as a kind of savior of the Italian tradition.

In 1927, Zariski left Rome for the United States, where he was impressed not only by the discovery of the topological methods of Lefschetz but also by the great progress made possible by introducing into algebraic geometry the rigorous methods of commutative algebra developed by Emmy Noether and her school. "It was a pity," Zariski wrote, "that my Italian teachers never told me there was such a tremendous development of the algebra that is connected with algebraic geometry. I only discovered this much later, when I came to the United States" [Parikh, 1991, 36].

The subsequent development of Zariski's ideas was thus more and more removed from the Italian school, as evidenced by the publication of his influential treatise on algebraic surfaces during his first decade in the United States [Zariski, 1935]. In some sense, this work, together with

Remarks on the Relations between the Italian and American Schools of Algebraic Geometry in the First Decades of the 20th Century

Aldo Brigaglia[a] and Ciro Ciliberto[b]

[a]Dipartimento di Matematica, Università di Palermo, Palermo, Italy
[b]Dipartimento di Matematica, Università di Roma Tor Vergata, Rome, Italy

11

ABSTRACT

In this paper we give an overview of the interactions between Italian and American algebraic geometers during the first decades of the 20th century. We focus on three mathematicians—Julian L. Coolidge, Solomon Lefschetz, and Oscar Zariski—whose relations with the Italian school were quite intense. More generally, we discuss the importance of this influence in the development of algebraic geometry in the first half of the 20th century.

INTRODUCTION

This paper aims at giving an overview of the interactions between Italian and American algebraic geometers during the first three decades of the 20th century. It focuses on three mathematicians—Julian Lowell Coolidge, Solomon Lefschetz, and Oscar Zariski—who represent three different periods in the development of algebraic geometry.

The first period extends roughly from the end of the 19th century up to the beginning of World War I. It witnessed the opening up of new vistas in algebraic geometry by the Italian school of algebraic geometry: building the theory of surfaces, creating new methods, and sparking the imagination of young mathematicians internationally. Guido

Castelnuovo characterized the endeavor as "the exploration of a vast territory seen from a faraway peak" [Enriques, 1949, vii]. André Weil concurred; in his view, "vast territories were being opened up" [Weil, 1946, viii]. This thriving atmosphere deeply impressed many young mathematicians coming, as Coolidge did, from abroad. Thus, this first period marked an internationalization of the Italian school of algebraic geometry, with Coolidge becoming a member of it, despite his birthplace.

The second period begins with the end of World War I and roughly coincides with both Lefschetz's first trip to Italy in 1920 and Zariski's first extended stay in Italy (1921 to 1927). During this second period, these two young mathematicians—coming from and going to the United States—were deeply influenced by the geometrical intuition and impressive results of the Italians as well as by what Zariski termed their "elegant style" [Parikh, 1991, 60]. However, both Lefschetz and Zariski soon became aware that new, more effective tools were required to overcome some of the major difficulties that confronted algebraic geometry—in particular, its need for sounder foundations. They found those tools in the development of topological and algebraic methods. Zariski especially became one of the main promoters of the systematic application to algebraic geometry of the techniques of commutative algebra, an idea refined and, to a large extent, created by the German school of Emmy Noether, Emil Artin, and Bartel van der Waerden. Their Italian mentors did not accept and, to some extent, did not even understand the need for the kind of deep transformations Lefschetz and Zariski effected. Indeed, by the end of the 1920s, the Italians' former students had begun to give birth to a radically new mathematical school.

The year 1935, the date of the publication of Zariski's book on algebraic surfaces, may be said to mark the opening of the third period. Even though his book did not put effectively to use the new algebraic methods that Zariski was developing in those years, it took a decidedly new stance toward such important problems as rigor in proofs and foundations. Moreover, this third period marked the gradual foundation by the former student in Italy, Zariski, of a new approach to both algebraic geometry in particular, and mathematics in general, with Zariski as its chief representative. The growing success of Zariski's school (and of André Weil some years later) in attacking the problems of classical algebraic geometry with the tools of the new "abstract" and "structural" algebra was of paramount importance in defining the new mathematical style known as "bourbachisme." Even if he personally continued to be tied to (and fascinated by) pure geometry, Zariski has been rightly considered the one who cut "the umbilical cord" [Lang, 1959, 245] that bound modern to classical algebraic geometry.[1]

Historians have tended to draw attention to Zariski's role in building the so-called new mathematics,[2] but, in doing so, they have almost always completely ignored the links between him and the Italian school. Moreover, they have often constructed a genealogy that goes directly from Dedekind and Weber to Zariski and Weil through Noether's school. Here, we argue that this interpretation is incorrect. The real aim of mathematicians such as Lefschetz and Zariski was not to contrast Italian approaches to geometry with the German school of algebra but to merge their methods. New mathematical methods were developed to overcome the difficulties of classical geometry but with the aim, as Weil put it, "to finish, in harmony with the portions already existing, what has been left undone" [Weil, 1946, viii].

JULIAN COOLIDGE IN TURIN

Julian Coolidge (1873–1955), a young American mathematician, traveled to Europe in 1903 to complete his studies. He went first to Paris, then to Griefswald, where he studied with Gerhard Kowalewski and Eduard Study, next to Turin, and finally to Bonn, where Study had moved and where Coolidge completed his doctoral thesis (see Struik [1955]).

During the academic year 1903–1904, Coolidge attended the lectures in higher geometry given by Corrado Segre at the University of Turin on applications of Abelian integrals to geometry. In some sense, Segre had headed and maintained the school of geometry that Luigi Cremona had established in 1860. Among Segre's students (in an extended sense of the term) were the leading Italian geometers of the early years of the

20th century: Castelnuovo, Federigo Enriques, and Francesco Severi. Through their works, the Italian school of algebraic geometry developed rapidly and enjoyed a growing international reputation.

At the third International Congress of Mathematicians, held in Heidelberg in the summer of 1904, Segre delivered a general lecture, in which he clearly expressed his and his students' research program [Segre, 1904]. It was the same research program to which Coolidge had been exposed during the academic year 1903–1904. Segre's course incorporated a short presentation of algebraic curves and many suggestions about the use of complex analysis in geometry.[3] It was also linked with several scattered ideas on these topics that he had begun to publish as early as 1889 Segre, 1889–1890a, Segre, 1889–1890b, Segre, 1889–1890c, Segre, 1890–1891 and Segre, 1891 as well as with some ideas of Study.[4]

The mathematics Coolidge became acquainted with in Segre's lectures, as well as his whole Italian experience, deeply impressed him. In 1904, he wrote a paper in which he detailed his impressions. "The pleasantest feature of the life of an advanced student in Italy," he wrote, "will be the personal contact with the professor in whose subject he is especially interested. Seminars do not exist, nor have I seen traces of students' mathematical clubs, but the relation between teacher and pupil is most direct and most helpful" [Coolidge, 1904c, 13]. Moreover, he continued, "Americans are sure to find lectures on subjects that will interest them, and they will have the French rather than the German standard of clearness and elegance. They will also be struck by the eclecticism of the instructor, for Italian mathematicians read widely. I remember being impressed at the beginning of one course of lectures by the fact that the professor put down, as principal works of reference, books in four different languages, and remarked that those of his hearers who could not read English, French, and German must certainly make up the deficiency in the course of the year" [Coolidge, 1904c, 13].

Coolidge had clearly found in Turin a thriving atmosphere. He was fully immersed in the international ambience around Segre, which included Segre's former student, Gino Fano, and in which English,

French, and German were commonly and widely known. Moreover, given that Fano had spent some time in Göttingen with Felix Klein, Coolidge's "direct and most helpful" relations with his teachers also included links to and contacts with the German school. Thus, it is not surprising that the European country in which he settled to complete his doctoral thesis [Coolidge, 1904a] was Germany. In particular, he went to Bonn, where Study's intellectual interests tightly interlaced with those of Fano and Segre.

Like Segre and Study, Coolidge had deep interests in the geometrical interpretation of complex numbers and complex functions and, so, in non-Euclidean geometry, an area in which he published often.[5] He also worked in algebraic geometry in the classical sense of the Italian school [Coolidge, 1917]. Significantly, his important book [Coolidge, 1931] was dedicated "to Italian geometers, dead and alive (Ai Geometri Italiani, Morti, Viventi)" [Coolidge, 1931, 1]. Given these research interests, Coolidge may be considered a representative of the classical Italian school of algebraic geometry.

In this sense, the adjective "Italian" must, however, be considered more a mark of a scientific working style than an indication of a national school [Brigaglia, 2001]. In fact, at the same time in which Coolidge was attending Segre's lectures, the two best known Italian geometers, Castelnuovo and Enriques, worked to convey their methods and main results—their scientific working style—to an international audience.[6] The conflation of the notions of general "working style" and "national school" comes through clearly in the thoroughly typical characterization given by David Mumford. "The Italian school, and notably Severi, Todd, Eger, and B. Segre," he wrote, "developed a general theory of Chern classes in the algebraic case" [Zariski, 1971, 74]. For Mumford, then, the word "school" referred not to a strictly national community but to a more general concept that encompassed, among other individuals, the Belgian Lucien Godeaux and the American Julian Coolidge.

The justification for adopting this more general understanding of the phrase "the Italian school of algebraic geometry" is borne out, moreover, by Coolidge's personal experiences. When he returned to Italy in 1927, he was received as a member of the school. Besides renewing his personal friendships and mathematical relations with his old friends, Castelnuovo, Enriques, and Severi, he gave a talk that was published in the journal of the Roman Mathematical Seminar [Coolidge, 1928]. He was also in the process of preparing the above-mentioned treatise on algebraic curves that he dedicated to his Italian confrères. In 1927, then, the hegemony of Rome—the Princeton of the 1920s, as Gian Carlo Rota [1988] called it—in the realm of algebraic geometry seemed to be firmly established, but times were radically changing.

SOLOMON LEFSCHETZ: THE NEED FOR NEW TOOLS

In 1920, the Russian-born Solomon Lefschetz came from the United States to Europe—and mainly to France and Italy—to spend his sabbatical year and to advance in his mathematical training. A year later, Oscar (then still Ascher) Zariski went to Rome from Russia. These two events, while indicative of the increasing reputation of Italian algebraic geometers, were full of important consequences for the relations between Italy and the United States relative to research in algebraic geometry.

At the time of his visit to Rome, Lefschetz had already gained an international reputation in mathematics as the winner of the important Bordin Prize of the Paris Académie des Sciences [Lefschetz, 1921a]. This paper, largely inspired by the Italian geometrical school, coherently developed one of the themes preferred by the Italians, namely, so-called hyperelliptic surfaces and Abelian varieties. Four Italians, in fact, had already won the same prize for results on the same subject: Enriques and Severi in 1908 Enriques and Severi, 1909 and Enriques and Severi, 1910 and Giuseppe Bagnera and Michele De Franchis in 1910Bagnera and De Franchis, 1910 and Ciliberto and Sernesi, 1991. In his work, Lefschetz was influenced not only by the ideas of these four Italians, but also by recent papers of Gaetano Scorza [1916] and Carlo Rosati [1915].

Lefschetz never denied his indebtedness to Italian geometers, and during the fall of 1920 and the winter of 1921, he had many talks with Scorza and, above all, with Castelnuovo. Lefschetz paid a warm tribute to Scorza and his work in the spring of 1921 in the Hadamard Lectures he gave at the Collège de France. Lefschetz was pleased to "recall that in the course of certain investigations, his [Scorza's] method whose elegance is beyond doubt, was of considerable help to me. Therefore, I believe that I am doing useful work by summarizing here the main features of his work" [Lefschetz, 1923, 120]. [7]

Yet, while deeply influenced by the Italian geometers, Lefschetz clearly perceived the need to renew the theory, to endow it with new and more powerful tools. He thus noted with surprise that the Italian geometers had not even tried to make use of topological tools. "Other phases of the theory were investigated in Italy and to some extent also in France, receiving an especially powerful impetus at the hands of Castelnuovo, Enriques and Severi," he wrote. "It is however a rather remarkable fact that in their work topological considerations are all but absent, practically never going beyond the study of linear cycles" [Lefschetz, 1921a; in Lefschetz, 1971, 42]. In that context he formulated his celebrated research program in a very short, but highly effective, sentence: "To plant the harpoon of topology in the whale body of algebraic geometry" [Lefschetz, 1968; in Lefschetz, 1971, 13]. In this way, he indicated a way to develop the Italian geometrical school (in the above-mentioned sense) starting from inside it, but pointing to new methods and new mathematical tools.

Italian mathematicians, fully confident in the effectiveness of their traditional projective methods, were scarcely aware of this, although Rosati, Scorza, and most especially Castelnuovo were exceptions. As is well known, Castelnuovo had essentially stopped publishing in algebraic geometry in 1906 [Brigaglia and Ciliberto, 1995], owing mainly to his sense of uneasiness in the face of the growing difficulties encountered in trying to extend the Italian methods (which had been so successful in the classification of surfaces) to the study of varieties of higher dimensions. In the early 1920s, though, Castelnuovo seemed to find refreshing his close contact with Lefschetz. He vividly recalled that "[t]he frequent conversations with Solomon Lefschetz during the winter of 1921 … induced me to publish a part of some extensive research on Abelian functions that I had begun several years before with the purpose of clarifying and extending the beautiful results of G. Humbert on singular hyperelliptic functions" [Castelnuovo, 1937, 549].[8] Castelnuovo published his first paper in 15 years on algebraic geometry in 1921 [Castelnuovo, 1921]; it appeared in the journal of the Accademia dei Lincei, together with another one by Lefschetz [1921b].

Obviously, this was not at all coincidental. The two articles were tightly integrated and presented, respectively, the analytical (Castelnuovo) and the topological (Lefschetz) aspects of the theory of Abelian functions. As Castelnuovo later recalled, "[a]nother push to the publication of this memoir came from the applications to irrational series of divisors on a curve, a subject which had been called to my attention more than once and to that of the young people who had continued the journey I had begun. This subject was then taken up by C. Rosati, who had already made valid contributions to it in earlier memoirs. With the premature deaths of R. Torelli and C. Rosati, research on irrational series slowed" [Castelnuovo, 1937, 549; our emphasis]. [9]

Ruggero Torelli (1884–1915), one of the most gifted young Italian algebraic geometers of the time, had been tragically killed during the First World War, while Carlo Rosati (1879–1929) passed away before the 1920s were out. Castelnuovo seemingly saw in the two Italians, as well as in their contemporary Lefschetz, representatives of a younger generation of researchers who, he hoped, could transform and develop Italian methods without overturning them. History did not bear out these hopes, however.

If Lefschetz served to join Italian and American algebraic geometers, he also linked Italian and American algebraists. In Italy, Gaetano Scorza had seen clearly the connections between Abelian varieties and their endomorphisms (i.e., Riemann matrices and their Hurwitz relations) and the modern theory of algebras, and he had published an influential paper on this topic [Scorza, 1921] that greatly impressed Lefschetz. In the United States, on the other hand, Adrian Albert was engaged in important research on the abstract theory of algebras when Lefschetz attended to one of his talks on the subject in 1928. According to Nathan Jacobson, "Lefschetz apparently sensed that here was a brilliant young algebraist whose interests and power made him ideally suited to attack his [Lefschetz's] problem" [Jacobson, 1974, 1076].

4. E.S. Allman, J.A. Rhodes, Phylogenetics, Modeling and Simulation of Biological Networks, Proc. Sympos. Appl. Math., Amer. Math. Soc., Providence, RI (2007), to appear.
5. M. Casanellas, S. Sullivant, The strand symmetric model, in: L. Pachter, B. Sturmfels (Eds.), Algebraic Statistics for Computational Biology, Cambridge University Press, Cambridge, 2005, pp. 305–321.
6. M.V. Catalisano, A.V. Geramita, A. Gimigliano, Ranks of tensors, secant varieties of Segre varieties and fat points, Linear Algebra Appl. 355 (2002) 263–285.
7. J.A. Cavender, J. Felsenstein, Invariants of phylogenies in a simple case with discrete states, J. Class 4 (1987) 57–71.
8. D. Cox, J. Little, D. O'Shea, Ideals, Varieties, and Algorithms, Undergraduate Texts in Mathematics, Springer, New York, 1997.
9. D. Cox, J. Little, D. O'Shea, Using Algebraic Geometry, Springer, New York, 1998.
10. D. Eisenbud, Commutative Algebra with a View Toward Algebraic Geometry, Graduate Texts in Mathematics, Springer, New York, 1994.
11. N. Eriksson, Toric ideal of homogeneous phylogenetic models, in: ISSAC 2004, ACM, New York, 2004, pp. 149–154.
12. N. Eriksson, K. Ranestad, B. Sturmfels, S. Sullivant, Phylogenetic algebraic geometry, in: C. Ciliberto, A.V. Geramita, B. Harbourne, R.M. Miro-Roig, K. Ranestad (Eds.), Projective Varieties with Unexpected Properties: A volume in Memory of Giuseppe Veronese; Proceedings of the international Conference 'Varieties with Unexpected Properties', Siena, Italy, June 8–13, 2004, Walter de Gruyter, 2005, pp. 177–197.
13. S.N. Evans, T.P. Speed, Invariants of some probability models used in phylogenetic inference, Ann. Statist. 21 (1) (1993) 355–377.
14. J. Felsentein, Inferring phylogenies, Sinauer Associates, Inc., Sunderland, 2003.
15. V. Ferretti, D. Sankoff, The empirical discovery of phylogenetic invariants, Adv. Appl. Probab. 25 (2) (1993) 290–302.
16. V. Ferretti, D. Sankoff, Phylogenetic invariants for more general evolutionary models, J. Theor. Biol. 173 (1995) 147–162.
17. L.D. Garcia, M. Stillman, B. Sturmfels, Algebraic geometry of Bayesian network, J. Symbolic. Comput. 39 (2005) 331–355.
18. J. Harris, Algebraic Geometry: A First course, Graduate Texts in Mathematics, vol. 133, Springer, Berlin, 1992.
19. R. Harthshorne, Algebraic Geometry, Graduate Texts in Mathematics, Springer, New York, 1977.
20. M.D. Hendy, The relationship between simple evolutionary tree models and observable sequence data, Syst. Zool. 38 (1989) 310–321.
21. M.D. Hendy, D. Penny, A framework for the quantitative study of evolutionary trees, Syst. Zool. 38 (1989) 297–309.
22. J.A. Lake, A rate independent technique for analysis of nucleic acid sequence: evolutionary parsimony, Mol. Biol. Evol. 4 (1987) 167–191.
23. J.M. Landsberg, L. Manivel, On the ideal of secant varieties of Segre varieties, Found. Comput. Math. 4 (4) (2004) 397–422.

24. L. Pachter, B. Sturmfels, The Mathematics of Phylogenomics, preprint, 2005.
25. L. Pachter, B. Sturmfels (Eds.), Algebraic Statistics for Computational Biology, Cambridge University Press, Cambridge, 2005.
26. C. Semple, M. Steel, Phylogenetics, Oxford Lecture Series in Mathematics and its Applications, vol. 24, Oxford University Press, Oxford, 2003.
27. M. Steel, Recovering a tree from the leaf colourations it generates under a Markov model, Appl. Math. Lett. 7 (2) (1994) 19–24.
28. B. Sturmfels, S. Sullivant, Toric ideals of phylogenetic invariants, J. Comput. Biol. 12 (2) (2005) 204–228.

CITATION

Cristiano Bocci, Dipartimento di Matematica, Università di Milano, Via Cesare Saldini 50, 20133 Milano, italy, topics on phylogenetic algebraic geometry, doi:10.1016/j.exmath.2007.02.001.

Lefschetz thus served to move the American algebraic school in the same direction Scorza and Rosati had gone in Italy, namely, toward an exploration of endomorphisms of Abelian varieties, i.e., multiplication of Riemann matrices, in a purely algebraic, as opposed to topological, way. Although Albert moved from pure algebra to algebraic geometry and Scorza moved from algebraic geometry to pure algebra, the two approaches they represented coalesced for a decade. Albert made it quite clear: "The structure of impure Riemann matrices was shown by G. Scorza to be expressible in terms of the structure of its pure sub-matrices, and the case where D [the multiplication algebra of a Riemann matrix] is a commutative algebra was discussed at length by S. Lefschetz. No further really important work in this direction was made until that of a recent paper by C. Rosati. … We give a rigorous algebraic proof of Rosati's main results and obtain his theorems, not in terms of unknown sub-algebras of given algebras B but actually in terms of the central field of B and its rank function" [Albert, 1931, 1]. Albert's work linked up with Scorza's during the 1930s. By that time, however, the Italian school of algebraic geometry was no longer capable of actively following the new developments spurred by Albert's works.

OSCAR ZARISKI: FROM UNEASINESS TO CRITICISM

When the 22-year-old Oscar Zariski came from Russia to Rome in 1921, Lefschetz was renewing his Italian connections.[10] Zariski remained in Rome until 1927, studying algebraic geometry mostly under Castelnuovo's supervision and learning both scientific and life lessons from his long and frequent conversations with Federigo Enriques. Zariski may thus be considered a student of both Castelnuovo and Enriques. His mathematical "style" and the methods he used differed little from those of the most brilliant of his Italian mathematical contemporaries—Luigi Campedelli, Oscar Chisini, Fabio Conforto, and many others.

By the close of the 1920s, however, a major shift occurred. While Chisini, Campedelli, and later Beniamino Segre and others did not feel the need—and were not encouraged by their teachers—to renovate the methods and ideas inherited from their masters, Zariski felt increasingly uneasy working in that tradition. In 1928, still considering himself a close adherent to Castelnuovo's methods, he attended the International Congress of Mathematicians held in Bologna. Writing to his wife, he gloried in his association with his mentor's mathematical thought: "Today Castelnuovo gave a lecture: it was the best so far and it made a great impression on everyone, both for its content and for its elegant style. It was a real work of art. He did me the greatest honour of interrupting his lecture at a certain point … in order to announce to the audience my upcoming communication to the Congress, in which, according to him, I have made an important step toward the solution of a fundamental problem which is still unresolved … you can understand how significant a sign of recognition it was" [Parikh, 1991, 60].

But 1928 also witnessed the first symptoms of Zariski's uneasiness with the Italian approach to algebraic geometry. It was apparently Castelnuovo himself who sparked this unease. He reputedly declared that "[t]he methods of the Italian school have reached a dead end and are inadequate for further progress in the field of algebraic geometry" [Parikh, 1991, 36]. If Castelnuovo did, indeed, make this pronouncement, it is easy to understand why he suggested that his young student should go further in his studies and "explore the work of Solomon Lefschetz" [Parikh, 1991, 36]. In the second half of the 1920s, Castelnuovo thus viewed the American approach as a kind of savior of the Italian tradition.

In 1927, Zariski left Rome for the United States, where he was impressed not only by the discovery of the topological methods of Lefschetz but also by the great progress made possible by introducing into algebraic geometry the rigorous methods of commutative algebra developed by Emmy Noether and her school. "It was a pity," Zariski wrote, "that my Italian teachers never told me there was such a tremendous development of the algebra that is connected with algebraic geometry. I only discovered this much later, when I came to the United States" [Parikh, 1991, 36].

The subsequent development of Zariski's ideas was thus more and more removed from the Italian school, as evidenced by the publication of his influential treatise on algebraic surfaces during his first decade in the United States [Zariski, 1935]. In some sense, this work, together with van der Waerden's Moderne Algebra, opened up entirely new developments in modern mathematics. Zariski's book, in particular, showed in several instances the fruitfulness of the insertion of modern topological (Lefschetz) and algebraic (Emmy Noether) methods into Italian algebraic geometry. Thus, it represented a further evolution of Lefschetz's program of "plant[ing] the harpoon of algebraic topology into the body of the whale of the algebraic geometry." As Serge Lang put it, the "umbilical cord" that had tied Zariski to his Italian teachers had begun to be cut in 1935 [Lang, 1959, 245]. Although Zariski's working style was, by then, deeply different from that of his Italian contemporaries, it would be a mistake to think that this development initially took place in opposition to the methods of Enriques; as noted above, it had been Castelnuovo who had encouraged some sort of shift. Only gradually did the declining Italian school come into conflict with the more advanced international tendencies that Zariski represented so well.

Here, we need not go more deeply into these problems, but it may be useful to note that just three years before Zariski published his treatise, another book on the same subject by Enriques and his student, Luigi Campedelli, appeared [Enriques and Campedelli, 1932]. A brief comparison of the two texts highlights the different points of view of their respective authors. Consider, for example, their stances on the resolution of singularities. Enriques and Campedelli asserted that the fact "[t]hat, effectively, the indicated procedure of transformations ends in the complete resolution of all singularities, has been rigorously proved by B. Levi and O. Chisini" [Enriques and Campedelli, 1932, 11]. [11] Zariski had a very different read on the same issue. In his view, "[t]he proofs of these theorems are very elaborate and involve a mass of details which it would be impossible to reproduce in a condensed form. It is important, however, to bear in mind that in the theory of singularities the details of the proofs acquire a special importance and make all the difference between theorems which are rigorously proved and those which are only rendered highly plausible" [Zariski, 1935, 18]. The radically different ways in which Enriques and Campedelli, on the one hand, and Zariski, on the other, viewed the "details of the proofs" underscores the opposition between the two different methods in algebraic geometry that had evolved (for more detailed comments on Zariski's book, see Brigaglia and Ciliberto [1995, §4.4]).

Zariski, however, never forgot—and continued to appreciate—the tradition in which he had been raised. "[A]lthough [he] recognized the power of sheaf theory, he felt in the last analysis it only reworked and made more intelligible the ideas inherent in the Italian approach" [Parikh, 1991, 158]. By the 1950s, then, relations between the Italian and American schools were suspended between a sense of continuity and of rupture.

Algebraic geometry, developing from complex algebraic geometry to algebraic geometry on an arbitraryfield, needed reformulation in terms of entirely new methods, which had to be independent of the characteristic of the field. While Zariski found his way in a new direction—even if his "Italian teachers never told" him of modern algebra—his Italian colleagues could not do the same. They were trapped in the stuffy academic atmosphere that prevailed in Italian universities. Many young Italian geometers of the new generation thus intensely felt the need to sever their ties with a tradition that had begun to seem to them too conservative. They found a great source of inspiration in the American school. The flux of the movement of ideas now moved from the American to the Italian school. The direction of influence was reversed.

From an historical point of view, algebraic geometers began in the early 1980s to rethink the links between the modern and the classical way of doing mathematics. We conclude with a quotation from David Munford that Castelnuovo might well have appreciated: "The 20th century has been, until recently, an era of "modern mathematics" in a sense quite parallel to "modern art" or "modern architecture" or "modern music." That is to say, it turned to an analysis of abstraction, it glorified purity and tried to simplify its results until the roots of each idea were manifest. ... Now the trend has reversed: post-modern mathematics is quite different and has reintroduced the love of the baroque» [Parikh, 1991, xxvii].

CONCLUSIONS

A critical reader could argue that the present paper does not contain anything new. In fact, in a sense, that is true. As we pointed out in the Introduction, however, our aim here has been to stress—rather than to illustrate in any detail—the influence that the Italian school of algebraic geometry had on the development of the ideas, and even on the education, of Coolidge, Lefschetz, and Zariski. While Coolidge marks the period of full glory of the Italian school, Lefschetz and Zariski had rather different experiences. Lefschetz's education was topological and analytical rather than algebraic and geometrical. His interaction with the Italian masters, though extensive and very appreciative, was thus more an encounter than a real sharing of ideas and methods, except, perhaps, for Rosati and Scorza, who were nevertheless rather isolated inside the Italian school. Zariski, on the other hand, though brought up as a student of the school, soon felt the need for an enlargement of his cultural and even geographical horizons. As we have tried to indicate, however, the break with tradition and the personal polemics which characterized the birth of so-called modern algebraic geometry, in contradistinction to classical Italian algebraic geometry, only came later in the course of the 20th century (see Brigaglia and Ciliberto [1995] for details). At the beginning, Zariski, as well as Lefschetz and the older Coolidge, felt deeply linked to the Italian tradition, probably more than with any other. In a certain sense, we can speak not of a break with a tradition, but rather of a natural evolution of methods and ideas that Zariski rightly promoted as a former member of that school and as the natural heir to its treasures. To put forward this alternative interpretation was the main purpose of this note. Of course, we might also have raised the interesting question: why did this natural evolution not take place inside the Italian school itself? This wider issue has, in part, been treated in Brigaglia and Ciliberto [1995] and goes well beyond the scope of the present paper.

ACKNOWLEDGMENTS

The authors warmly thank Prof. Karen Parshall for her careful revision of the English form of the first version of the present paper.

REFERENCES

1. Albert, A., 1931. The structure of pure Riemann matrices with non-commutative multiplication algebras. Rend. Circ. Mat. Palermo 55, 1–59.
2. Bagnera, G., De Franchis, M., 1910. Le nombre ⊠0 de M. Picard pour les surfaces hyperelliptiques et pour les surfaces irrégulières de genre zéro. Rend. Circ. Mat. Palermo 30, 185–238.
3. Brigaglia, A., 2001. The creation and persistence of national schools: The case of Italian algebraic geometry. In: Bottazzini, U., Dahan, A. (Eds.), Changing Images in Mathematics. Routledge, London, pp. 187–206. A. Brigaglia, C. Ciliberto / Historia Mathematica 31 (2004) 310–319 319
4. Brigaglia, A., Ciliberto, C., 1995. Italian Algebraic Geometry between the Two World Wars. In: Queen's Papers in Pure and Applied Mathematics, vol. 100. Queen's University, Kingston.
5. Castelnuovo, G., 1921. Sulle funzioni Abeliane. Rend. Accad. Lincei (5) 30, 50–55; 99–103; 195–200; 355–359.
6. Castelnuovo, G., 1937. Memorie Scelte. Zanichelli, Bologna.
7. Ciliberto, C., Sernesi, E., 1991. Some aspects of the scientific activity of Michele De Franchis. In: De Franchis, Michele (Ed.), Opere. Suppl. Rend. Circolo Mat. Palermo.
8. Coolidge, J.L., 1904a. Die dual-projektive Geometrie im elliptischen und sphärischen Raume. Dissertation, Bonn.
9. Coolidge, J.L., 1904b. Les congruences isotropes qui servent à représenter les fonctions d'une variable complexe. In: Atti Accad. Scienze Torino.
10. Coolidge, J.L., 1904c. The opportunity for mathematical study in Italy. Bull. Amer. Math. Soc. 11, 9–17.
11. Coolidge, J.L., 1908. The Elements of Non Euclidean Geometry. Clarendon, Oxford.
12. Coolidge, J.L., 1913. A study of the circle cross. Trans. Amer. Math. Soc. 14, 149–174.
13. Coolidge, J.L., 1916. A Treatise on the Circle and Sphere. Clarendon, Oxford.
14. Coolidge, J.L., 1917. The characteristic numbers of a real algebraic plane curve. Rend. Circ. Mat. Palermo 42, 260–266.
15. Coolidge, J.L., 1928. Questioni di geometria nel campo complesso. Rend. Sem. Mat. Roma 2, 24–29.
16. Coolidge, J.L., 1931. A Treatise on Algebraic Plane Curves. Clarendon, Oxford.
17. Corry, L., 1996. Modern Algebra and the Rise of Mathematical Structures. Birkhäuser, Basel.
18. Dieudonné, J., 1974. Cours de Géométrie Algébrique. Presses Universitaires de France, Paris.
19. Enriques, F., 1949. Le superficie algebriche. Edited by G. Castelnuovo, A. Franchetta, and G. Pompilj. Zanichelli, Bologna.
20. Enriques, F., Campedelli, L., 1932. Lezioni sulla teoria delle superficie algebriche. CEDAM, Padova.

21. Enriques, F., Severi, F., 1909. Mémoire sur les surfaces hyperelliptiques. Acta Math. 32, 283–392.
22. Enriques, F., Severi, F., 1910. Mémoire sur les surfaces hyperelliptiques. Acta Math. 33, 321–403.
23. Giacardi, L. (Ed.), 2002. I Quaderni di Corrado Segre. Università di Torino. CD produced by the Dipartimento di Matematica.
24. Jacobson, N., 1974. Abraham Adrian Albert. Bull. Amer. Math. Soc. 80, 1074–1100.
25. Lang, S., 1959. Abelian Varieties. Interscience, New York.
26. Lefschetz, S., 1921a. On certain numerical invariants of algebraic varieties with application to abelian varieties. Trans. Amer.
27. Math. Soc. 22, 327–428.
28. Lefschetz, S., 1921b. Sur le théorème d'existence des fonctions abéliennes. Rend. Accad. Lincei (5) 30, 48–50.
29. Lefschetz, S., 1923. Progrès récents dans la théorie des fonctions abéliennes. Bull. Sci. Math. 47, 120–128.
30. Lefschetz, S., 1968. A page of mathematical autobiography. Bull. Amer. Math. Soc. 74, 854–879.
31. Lefschetz, S., 1971. Selected Papers. Chelsea, New York.
32. Parikh, C., 1991. The Unreal Life of Oscar Zariski. Academic Press, New York.
33. Picard, E., Simart, G., 1906. Théorie des fonctions algébriques de deux variables indépendants, vol. II. Gauthier–Villars, Paris.
34. Rosati, C., 1915. Sulle corrispondenze tra i punti di una curva algebrica e, in particolare, tra i punti di una curva di genere due. Ann. Mat. Pura Appl. 25, 1–32.
35. Rota, G.C., 1988. Fine Hall in its golden age. In: A Century of Mathematics in America. American Mathematical Society, Providence, RI, pp. 223–236.
36. Scorza, G., 1916. Intorno alla teoria generale delle matrici di Riemann e ad alcune sue applicazioni. Rend. Circ. Mat. Palermo 41, 263–380.
37. Scorza, G., 1921. Le algebre di ordine qualunque e le matrici di Riemann. Rend. Circ. Mat. Palermo 45, 1–204.
38. Segre, C., 1889–1890a. UN nuovo campo di ricerche geometriche. Atti Accad. Scienze Torino 25, 180–205.
39. Segre, C., 1889–1890b. UN nuovo campo di ricerche geometriche. Atti Accad. Scienze Torino 25, 290–317.
40. Segre, C., 1889–1890c. UN nuovo campo di ricerche geometriche. Atti Accad. Scienze Torino 25, 376–396.
41. Segre, C., 1890–1891. UN nuovo campo di ricerche geometriche. Atti Accad. Scienze Torino 26, 35–71.
42. Segre, C., 1891. Le rappresentazioni reali delle forme complesse e gli enti iperalgebrici. Math. Ann. 40, 413–467.
43. Segre, C., 1904. La geometria d'oggidì e i suoi legami con l'analisi. In: Verhandlungen des 3en Int. Math-Kongress in Heidelberg, pp. 109–120.
44. Struik, D., 1955. Obituary: James Lowell Coolidge. Amer. Math. Monthly 62, 669–682.

45. Weil, A., 1946. Foundations of Algebraic Geometry. American Mathematical Society, Providence, RI.
46. Zariski, O., 1935. Algebraic Surfaces. Springer-Verlag, Berlin.
47. Zariski, O., 1971. Algebraic Surfaces, second ed. Springer-Verlag, New York.

CITATION

Aldo Brigaglia, Ciro Ciliberto, Remarks on the relations between the Italian and American schools of algebraic geometry in the first decades of the 20th century, Historia Mathematica, Volume 31, Issue 3, August 2004, Pages 310-319, ISSN 0315-0860, http://dx.doi.org/10.1016/j.hm.2003.09.003.

Index